This is a book for our time - a self-help book to inspire healing and spiritual growth for all of humanity. Some would say we are heading for self-destruction; that our history of lethal violence towards our own kind has programmed us to be fearful, suspicious and revengeful; and that, despite our modern age, this programming now threatens our well-being on a global scale never witnessed before.

But does it have to be this way?

Can we not consciously evolve into a different kind of human, from Homo sapiens to Homo spiritus?

This motivational book guides us in six steps towards that possibility. It is not an easy journey; it means reversing loops of bio-emotional stress responses which we do every day, unconsciously, all our lives. It is about changing the human mind-set and healing the past and if we succeed, and we can, then there is real hope for the future.

The author has been a health practitioner since the early 90's, specialising in a holistic approach and over the years has undertaken numerous courses in health related modules, including a trip to China to observe Traditional Chinese Medicine.

As well as being the founder/owner of Fountains Court in Scarborough, a hotel specialising in well-being, the author gives talks and seminars for corporate businesses, college students, medical students and special interest groups.

A Cordon Bleu chef and a one-time member of Mensa; with a degree in Humanities with Philosophy, plus a passion for writing, the author presents a unique book for readers of all ages who have within them the spark of Homo spiritus.

'We cannot change the past but we can heal it.'

HOMO SPIRITUS

A different kind of Human

HELEN MARRIOTT
BA, M.T.I, B.F.R.P, I.P.T.I.

Copyright © E. Helen Marriott – 2017
All rights reserved. No part of this publication may be reproduced, stored in a retrieval system, or transmitted, in any form or by any means, electronic, mechanical, photocopying, recording or otherwise, without the prior permission of the copyright owner and the publisher.

E. Helen Marriott asserts her right to be identified as the author of this work in accordance with the Copyright, Designs and Patents Act 1988.

Should we have failed to trace and acknowledge any copyright holder please contact the publisher who will make corrections in any future edition of this book.

This edition published in Great Britain in 2017 by

Farthings Publishing
8 Christine House
1 Avenue Victoria
SCARBOROUGH
YO11 2QB
UK
http://www.Farthings-Publishing.com
E-mail: queries@farthings.org.uk

ISBN 978-1-326-96369-9
May 2017 (f)

Cover graphic courtesy of Shutterstock

Contents

		Page
Introduction	We start from here	7
	Happiness	13
	A bit about the author	17
	Gratitude	18
	Acknowledgements	18
PART 1	The first 3 steps	21
	STEP 1 - Breathe better	22
	STEP 2 - Stress less	48
	STEP 3 - Fear not	70
PART 2	The next 3 steps	89
	STEP 4 - Be well	95
	STEP 5 - Live well	131
	STEP 6 - Love well	157
	Footnote	206
	Recommended reading	207
	Recommended therapies	208

INTRODUCTION

WE START FROM HERE

A lot of people would say that we – Home sapiens, are in a bit of a mess.

Despite our modern age we suffer stress, obesity, starvation, poverty, depression, anxiety, conflict, wars, corruption, terror ...

It seems like we are floundering in a sea of troubles ...

We're over-medicated, hospitals are full ...

Drugs and crime are widespread and our prisons are overcrowded.

If only things could be different ...

If only we could *feel* better ... feel less tired, less hopeless and less fearful.

If only we could all get *on* better with one another ... less prejudice, less greed, less anger.

If only we could walk out of this sea of troubles and find a better way forward.

At lot of people would say that's not possible, that as a species we are beyond hope, that it would take another Extinction Event – like a giant meteor or a catastrophic climate event or a nuclear war to wipe us out and for evolution to start again and maybe do better next time.

Hmm, but hang on, here's a thought ...

If evolution is the adaptation of life due to changes in the environment, what does that say about our part in that process?

There is a theory that we have evolved from knuckle-walking, tree-dwelling apes to upright-walking, prairie hunters - Homo erectus; followed by tool-makers – Homo habilis and finally to an animal with a large brain able to think, talk, count and reason and who developed a sense of Self with an awareness of a past, a present and even a future.

Homo sapiens had arrived.

However, being so clever and forward thinking, we reversed that loop of the environment shaping our evolution by changing the environment to suit *our* needs.

We moved from being nomadic hunters and foragers, living in small groups - connected *to* the land -and became settlers – claiming ownership *of* the land; building permanent dwellings that became villages, towns and cities.

We cut down forests, tamed nature and domesticated animals for food and tilled the earth to grow crops to help feed our growing populations.

We built roads, bridges; damned rivers.

And as civilisations developed so did trade routes, over land and sea. We bartered, invented money, weights and measures.

Economies developed as well as hierarchies, politics and legal systems.

We fought to adapt the environment to suit our needs and created our own living spaces and didn't we do well?

Homo sapiens are everywhere, from frozen ice-caps to scorched deserts; billions of us, and yet we still seemed to struggle to survive for we became victims of our own success in so many ways ...

Our settlements became centres for disease caused by poor sanitation and cross-infections like flu and typhoid from the animals that shared our lives.

Ruling Class systems developed creating fear, envy, suspicion and above all inequality, envy, greed, cruelty and subjugation. As they still do today.

Corruption, crime, land-grabbing and slavery became woven into the fabric of our existence as did the ever present cycles of war as we found more and more 'clever' ways to kill our own kind as we competed for just about everything.

It seems that despite being so clever we have our fingers permanently on the self-destruct button.

It seems that despite becoming masters of self-determination - the ability to make far-reaching choices - we are somehow advancing our own extinction.

It may indeed take another Extinction Event for natural evolution to create a different kind of human but supposing Homo sapiens *chose* to evolve into an even more advanced thinking-knowing human, could we do this?

Could we reverse more loops - the many cycles of cause and effect that govern our behaviour, our thinking, our existence?

It's amazing to think there was a time when we seriously believed that dunking an old woman in a pond would prove if she was a witch; if she drowned she wasn't a witch and if she didn't drown she was, and then we could burn her and purify her soul; whichever way the poor soul had an horrific death.

We now think that's a crazy idea.

Changing our mind-set is what we do. Right?

The Earth is not flat or supported by tortoises.

The Sun is not a god.

Sacrificing a human life to ensure a good harvest is just madness.

Slavery is never justifiable.

Women have as much right to self-determination as do men.

Such changes in how we think usually happens unconsciously as a slow awakening prompted by new ideas and philosophies, discoveries and inventions and as such can be a mixed-bag and is often met with huge resistance from those who hold power.

Such is the history of Homo sapiens.

So what about today?

In recent decades there has been a huge leap in our knowledge as Artificial Intelligence, and Quantum Physics reveal a rather weird hidden universe ... a universe that challenges all our previous notions of time and space and causality.

And then we have the extraordinary advancements in human biology, psychology and genetics.

With the discovery that our brains are not as hard-wired as we once thought, plus the idea that our genes have potentiality, i.e. behave much like quanta of energy rather than solid bits of matter, can we not deliberately set out to change how we think and react and hence how we behave?

Is it not time for another change of mind-set?

Some would say it is long overdue.

The question is what shape would it take?

Well, here's a thought:

What if Homo sapiens evolved into an aggressive, lethal human due to countless generations living in crowded, dangerous, war-faring societies?

This would, after all, be a natural bio-emotional evolutionary response to long-term threats of fear, famine, power-struggles and greed.

A response that is inevitably programmed into our genes and what's worse is that these genes continue to be strengthened because we continue to live this way.

It is a ghastly loop of cause and effect.

So, the thought is this - can we not purposefully reprogram our behaviour at the genetic level? Not by any kind of external engineering (heaven forbid!) but by adapting our response to external factors?

Much like evolution, in fact.

Surely we have the intellect and the willpower to do this or must we forever be the victims of our in-bred fear, greed and anger?

Can we not become a more peace-loving, caring, less fearful human being?

The answer is yes, we can, if we want.

But it ain't easy.

If it was we'd have done it thousands of years ago, but thousands of years ago we didn't have the knowledge we have today – of how our brain works, our psychologies, our hormones and all the rest of it.

Another thing in our favour is our amazing capacity to feel empathy and compassion; Homo sapiens frequently show a deeply caring nature.

Adolf Hitler, one of the most reviled persons in History and responsible for the slaughter of millions, supposedly loved his dogs, was a vegetarian and liked children.

So how do we solve this confusing duality in our behaviour, unbelievable cruelty on one hand and kindly compassion on the other?

How can we eradicate the cruelty and strengthen the compassion?

On the face of it, it seems an impossible task, so a change in mind-set is what is required and to do this we have to flip the coin ...

... What if there is no such thing as a Witch?

When we have this thought we begin to see things differently ... if there is no such thing as a Witch then why are we dunking old women in ponds?

Flip the coin – reverse the loop of habitual thinking.

So, in the case of our current situation it could go something like this ...

If we could only achieve World Peace we'd all feel much happier.

Flip to ...

If we all felt happier the world would be a more peaceful place.

Yeah, right!

Actually, it is right.

What we don't realise, or have forgotten, is that we have an enormous capacity to feel happy – our complex brains and our endocrine system (hormones) are massively equipped to give us that sense of joy which Homo sapiens feel as a *conscious awareness* that lifts our spirits.

(You may want to read that last paragraph again.)

Okay, while that little lot sinks in we'll stop a minute and ponder on that elusive word 'happy' ...

HAPPINESS

"Happiness, happiness, the greatest gift that I possess." (Song by comedian, Ken Dodd)

Comedians love to make people laugh, it's what they do and generally speaking, when we laugh it means we are enjoying a moment of happiness.

Happiness, though, is very much a subjective state of mind. A mountaineer would be very happy climbing a rocky mountain with ropes and pitons; I'd be terrified, miserable and probably cry.

Generally, though, we do not feel happy when we're ...

- Anxious
- Ill
- In debt
- Overweight
- Hungry
- Stressed
- Don't like what we do
- Don't like who we are
- Don't like who we're with
- Don't like where we live
- Don't feel loved
- Poor
- Fearful

And the list goes on ...

Feeling unhappy is our bio-emotional response to being stressed.

Conversely, feeling happy is our bio-emotional response to living a stress-free life, or at least to forgetting our stress for a while.

Humans are very clever at doing this; we have created all manner of distractions to help us forget how stressful life is. We read books; see a film; watch a comedian; play sports; get drunk ....

It seems we need these distractions, particularly those that make us laugh, in order to make life bearable ... hmmm.

Maybe this is one way of reversing the stress but it's always temporary. When the book is finished ... when the film ends ... when the bottle of wine is empty ... bam! We're back to reality

Equally obvious is the fact that we can't force ourselves to be happy. We may be able to pretend we are, and maybe fool others but we never fool ourselves.

We may pretend we're happy being two stone overweight but if someone had a magic wand to make us slim we'd leap at it.

But magic wands don't exist and going on diets just make us feel more miserable and vulnerable to failure.

Fairy godmothers, too, don't exist so no-one is going to pay our debts or give us a dream home.

So what is there to do? How does that list of being unhappy transform itself without a good dose of luck and magic?

It doesn't ... we do and we're the ones who work the magic.

Okay, we are now very close to taking that first not-so-easy step ...

Ready?

In order to change our current mind-set of fantasising about winning the lottery, meeting Mr or Mrs Right; waiting for stress to magically disappear ... we flip the coin:

If my life was stress-free I'd be happy

Flip to:

If I was happy my life would be stress-free

That's it?

In a nutshell, yes, but the path to peace and happiness is full of obstacles, we need a map ...

Mapping it out

To embark on any journey we need to make some kind of preparation before stepping out.

For the rise of Homo spiritus we need to understand what stress and unhappiness does to our physical and emotional well-being ... our body, mind and spirit.

Only then will we be in a position to change how we think, feel and react to things that seem beyond our control.

You see, to change our lives - our world - we have to consciously and deliberately reverse the bio-emotional loops that cause our stressful, miserable, angry, jealous, self-destructive behaviour.

Just imagine ... if we could succeed in consciously re-programming the potentiality of our genes that govern destructive human behaviour, think what a quantum leap in our evolution this could achieve!

This book explores that possibility and we need to start at the very beginning.

We need to start with who and what we are and examine the loops that for millions of years have chained us to our ancestral behaviour of fear and violence.

time to change.

know this and we can make that change if we choose.

starts and ends with each individual, you and me, until such time as a Critical Mass is reached and it is then that something quite remarkable can happen and Homo spiritus – a human with an awareness of joy, as yet unknown, can walk this astonishing planet.

What are we waiting for?

It's time to take that first step.

As with any new and radical idea, baby-steps are needed so we start with the basics. We are and always will be of a physical body and it is this body that needs very careful attention.

To understand how the 'machine' works – whatever make or model, helps us to prevent a break-down in the journey.

As for 'spirit', well that is a difficult word to define.

All cultures, ancient and modern have a grasp of what 'sprit' means ... not of the flesh - other worldly - the Essence of something - Universal Consciousness?

Or something that is precious, divine, beyond the confines of space and time?

Maybe.

This book is written in two parts but do be warned, it is not a good idea to skip to Part 2 until Part 1 has been digested. It is like a jelly waiting to set, if the jelly is turned out to soon it will not hold its shape and be difficult to swallow.

The book is written as lightly and un-academic as possible in the hope that anyone, teenagers especially, may read and enjoy.

Our bodies are extraordinarily intelligent and it's important to listen to what they are trying to tell us.

If we are not happy in our own bodies we are low in spirit and If we are not as fit and healthy as we could be, then it may difficult to reach our full potential, which is what Homo spiritus is all about.

So, we're ready to take that first step, but before we do

A bit about the author

The author was blessed from the start with a father who had the finest of minds, a mother with the warmest of hearts and a childhood filled with insects and snails; frogs, birds and small furry animals.

She was 'allowed' to be untidy and a daydreamer; to be an original thinker with an enquiring mind - an unconventional daughter - encouraged to follow her own star.

This was a wandering star which pulled her across oceans and continents, exploring the beauty of the natural world, different cultures, art and religions.

While still a young woman she witnessed an extraordinary event* which propelled her towards a deeper understanding of the Cosmos and the human mind.

This led to a degree in Philosophy with Humanities; a noble attempt at understanding Quantum physics and ultimately towards Medicine.

The body/mind connection had long been a fascination so it was Alternative and Complementary Medicine that became her chosen path.

Now, after thirty years of study and Practice the author shares her deeper understanding of who we are and her conclusion that we *can* change, if we chose to.

What surprised her was how comprehensive that change can be.

*this event will be detailed in a book called 'Three Worlds'.

Gratitude

With over six decades of being curious; adventurous; meeting people; listening; reading, learning; travelling and working in the field of Alternative Health - the knowledge gleaned and the ideas formulated in this book are due to all these things.

Having read so many books and articles it is impossible to single out any few for heartfelt appreciation although a suggested reading list is at the back of the book, including some for very young enquiring minds.

It is the same with people ... the strangers met through chance while travelling; tutors, colleagues, my own workshop students and in particular the many hundreds of clients and guests who have graced Fountains Court and shared with me their struggles and often remarkable stories.

Without their honesty, trust and courage Homo spiritus would not have materialised and I can tell you now, more than a few already walk this Earth.

The journey is far from over - it is barely started, and I look forward with a shining optimism to where the younger generations lead us.

The baton is passed on to you – take it, the future is yours.

Acknowledgements

Without my staff and colleagues who help to keep everything ship-shape as well as offering encouragement, suggestions and proof reading skills, this book may never have reached publication, particularly Kath Loughnan, Gilli Layton and Lesley Hodgson.

Clients and guests, too, played their part, particularly Sarah Dew and Monica Andersson whose enthusiasm for Homo spiritus carried me on a crest of a wave.

So too my amazing offspring, Sophie, Tom and Eddie; who, all their lives never complained when the seas got rough and who always seemed to trust in my destination. And also my mother, for being the legend she is and for waving the flag.

Then there's Mark, an anchor if ever one was needed; capable of feeding himself (and me) and who never complained when the Muse took-over and transported me far beyond the horizon.

Last but not least, for David, at Farthings Publishing – whose dedication and patience made the book a reality.

*

PART 1

THE FIRST 3 STEPS

STEP 1

BREATHE BETTER

Some cultures believe that spirit and breath is one and the same thing, thus, when a new born infant (human or animal) takes its first breath that is when 'spirit' enters its body, and likewise, when the spirit leaves the body that is the moment it takes its last breath.

It is certainly true that taking a breath is the first and the last thing we do and we all know that breathing – an exchange of gases - is vital for life on planet earth.

Okay, we need to get a bit technical here but stick with it; it's important and utterly beautiful.

All land mammals have an amazing muscle, called the diaphragm, situated under the rib-cage separating the chest from the abdomen in a nice and tidy fashion.

It is a flattish muscle and its ONLY function is to expand the rib cage so a vacuum is created inside the lungs causing air to be sucked in through the nose and down into the lungs which expand like balloons.

The diaphragm then relaxes - the lungs deflate and the air minus its oxygen (which has been absorbed into the blood), is puffed out through the nose.

And *that*, my loves, is the essence of breathing and, like all land mammals we do it unconsciously from the minute we are born.

The crying of a new born baby tells us this mechanism is working just fine and at the end of life the last thing that happens is we take a small ragged breath of air then as the diaphragm relaxes for the last time, we slowly let go of our fragile hold on planet Earth.

It is also the muscle that contracts quite painfully when we have hiccups. Other mammals can have hiccups, too.

However, with humans there are some mysterious, unique and extraordinary connections with breathing that is more than just the exchange of oxygen and carbon dioxide.

Even our ability to talk is dependent on breathing. We can only talk on the out-breath. Try breathing-in and talk at the same time ... no-can-do.

Humans would never have evolved speech without this amazing, often ignored, forgotten muscle.

It is also our chuckle muscle; it also heaves and shudders when we cry; it is at the very centre of our emotions, (that's a big clue for what's to come.)

First, we need to look at an extraordinary taken-for-granted human ability:

Ready?

We can choose how we breathe

Humans have freedom of choice; we have free-will which is probably why we're in the mess we're in.

Other animals behave according to their instincts and learned behaviour so *their* choices are limited.

In a nut-shell it looks like this:

INSTINCT drives the life of animals.

DESIRE drives the life of humans.

Now, you may wonder what all this has to do with breathing, well it's because there is a wider picture, a much wider picture connected with breathing than most of us ever realise.

Basically, we humans can control our breathing, consciously. We can choose how we breathe.

If I want to I can sit here and 'pant'.

I can hold my breath (not for ever because I'd pass-out).

I can even play around with my breathing – breathe in for the count of six, hold it for the count of four then breathe out for the count of four ... or whatever I choose.

No other animal can do this.

Because we can control our breathing we have the ability to sing, whistle ... anyone seen a dog or chimpanzee whistle? Ever seen a cat hold its breath?

All our other organs function beyond our will-power, they are controlled by our autonomic nervous system – we cannot, for example tell our heart to beat faster or our kidneys to filter slower.

Thankfully our breathing is also an autonomic function - if we had to concentrate every minute to breathe in-out we'd never do anything else.

AND YET we can over-ride that autonomic function.

How does this happen? It happens because, firstly, the diaphragm is a skeletal muscle, like our Pecs, ham-strings etc. Secondly, we can take wilful control of our skeletal muscles.

We can join a gym, yes? And learn how to use weights and do press-ups, (anyone see an ape do press-ups?)

However, muscles also react automatically to stress by bunching-up, preparing for flight, fight or freeze which is a perfectly normal biological response to danger for all animals and triggered by important hormones, without which life could not exist.

For Homo sapiens - a super-intelligent, knowing,

thinking kind of animal - things get a lot more complicated because between those two momentous occasions of our first and last breath things can go horribly wrong.

At an early age our breathing can start to dysfunction.

There are lots of reasons why we may breathe incorrectly - disease, pollution, genetics, injury, life-style, stress ...

The truth is, we should all be breathing from our tummies – babies and cats for example are wonderful tummy breathers, they are breathing correctly, from their diaphragms in a lovely relaxed manner.

From quite an early age, though, we humans start to breathe from our chests and if we're not breathing with our diaphragms we're not getting sufficient oxygen.

End of.

By the time Homo sapiens are teenagers we can start to become oxygen depleted and by the time we're in our thirties we can be fairly oxygen starved and this is very bad news indeed.

The problem is this, there is a finite supply of oxygen in our blood at any given moment and every cell in the body is screaming 'me, me me!' and when there isn't enough to go around ... well ... we're in trouble.

Why?

Because as everyone knows, our blood is pumped through the body by our beating heart, so when there's insufficient oxygen in the blood due to dysfunctional breathing, the poor heart has to work extra hard to try and get that limited supply of oxygen to every 'screaming' cell.

If that supply is less than it should be then we start on the slippery slope to serious health problems, such as:

High blood pressure, adrenal exhaustion, aching muscles, poor digestion, anxiety, sleeplessness, low immunity ... you name it ... it all starts and ends with the way we breathe.

And that's not all …

It is astonishing how many cubic centimetres of oxygen the body needs every second of the day and the **brain** is the most demanding organ of the lot, it demands one-quarter of all the oxygen in our body.

Have you ever rubbed your neck muscles or forehead when feeling tired, stressed or confused? That is a response to your brain demanding oxygen and this 'massaging' is an unconscious way of pushing blood to the front cerebral cortex – our 'thinking' brain.

When our brain has insufficient oxygen due to dysfunctional breathing we simply cannot think clearly, make decisions or cope with stress so the cycle becomes self-perpetuating.

This massaging is also an effort to relieve the soreness and stiffness from the neck and shoulder muscles that have responded to stress by bunching-up. This bunching-up impedes the flow of blood to the brain so we have yet another loop of cause and effect.

Ok, so humans, a knowing-thinking Being seem to have evolved a problem with their day-to-day breathing, so why has this happened?

Well, there is one outstanding cause for human dysfunctional breathing that sets us apart from all the other oxygen-breathing animals and it's linked to why we laugh and cry …

We are EMOTIONAL beings

Without doubt Homo sapiens are a very emotional species.

This doesn't mean that other animals don't experience emotions, anyone who has a pet or works with animals know this for a fact.

What makes us unique is that we are *conscious* of our emotions, of our personalities, of having desires and a free will which means we are constantly making plans for the future even if it's just what to eat for dinner.

All too often these plans can go wrong or back-fire and of course have an impact on others and *their* plans and desires.

All of this makes everything so much more complicated as we are constantly buffeted, influenced and affected by millions of 'decisions', good or bad, wise or unwise, kind or cruel made by ourselves, other individuals, groups, government or countries.

Hence, we are aware of feeling dozens of emotions - fear, anger, hatred, envy, excitement, joy, apathy, enthusiasm, sadness, and the less obvious mental states - indecision, nervousness, guilt, obsession, hilarity, misery, etc.

In fact, we are more or less 'ruled' by our emotions – victims, some might say, and what we find is this – our breathing muscle – our diaphragm, reacts to how we are feeling.

The diaphragm –linked to our solar plexus which is a mini-brain - 'listens' constantly to how we are feeling and our muscles and hormones react likewise with either a calm, happy, fear, flight, fight or freeze response, depending on the circumstances.

Picture the scene ...

It is a cold winter's evening; there's no moon and a wind-driven sleet fills the dark sky.

Phil can't wait to get home. He hurries along the street

from where he's parked his car, his head bent, a scarf wrapped around his neck and runs quickly down the steps to his basement flat.

A minute later he's in the front door, unwrapping the scarf from his neck and tweaking the thermostat on the wall.

Today is Phil's birthday, he's twenty-three. He picks up some mail from the floor and then hears a noise from down the hallway which is short and narrow with a door at the end that leads to the kitchen. The door is closed.

Holding his breath Phil listens, staring at the door. He listens hard.

Nothing.

Just the weather.

He lets out his breath and smiles as he recognises envelopes in his mother's and grandmothers' handwriting. His mobile phone rings – his father's face appears on the screen.

'Hi, Dad ... yeah, good thanks ... going to the rugby club later and have a bevy with the lads...Kate's coming down from Uni at the weekend, we'll celebrate then ...'

The sound of broken glass comes from the kitchen.

Phil catches his breath and freezes, staring at the door. 'Got to go, Dad, speak later.'

Sucking in more air he reaches for an umbrella propped against the wall. He is a big, strong bloke but feels better holding a sharp instrument that could give someone a nasty jab in the stomach.

Holding his breath he tip-toes towards the kitchen, gripping the umbrella like a bayonet. Grabbing the door handle he flings the door open and throws the light switch.

His kitchen is packed with burly young men who start to cheer and jeer and point at the umbrella in mock terror.

Phil puffs out his cheeks as the breath escapes from his

lungs and starts to cuss and laugh with the best of them.

In the middle of the floor is a large cardboard box wrapped with a red ribbon.

His rugby pals sing 'happy birthday' and tell him it's his birthday present and Phil has a good idea what it is.

Grinning he releases the ribbon, opens the flaps and looking into the box and sees Kate looking up at him, knees to her chin, eyes sparkling and very pink in the face.

'Get me out of here, I'm cooking!'

Laughing, Phil reaches in, puts his hands under her arms and lifts her out. Everyone cheers and whistles as Katy flings her arms around his neck and her legs around his waist.

'Baby, I'm so happy to see you, it's been weeks!' She says and starts to cry softly.

Phil twirls around and they kiss and are told to get a room and to be ready in half an hour because a table is booked at the local Tandoori.

So, lots of emotions in that little scene; with lots of different breathing techniques.

Sometimes our moods can become fixed, as in depression; or change in a flash as in a sudden burst of laughter or a bout of crying.

No other animal can laugh and cry the way we do.

As babies and toddlers we start to experience normal human emotions and by the time we are three years old we begin to know what we want and don't want.

We have temper-tantrums to get our own way - the clenched fists, the expanded rib-cage, the rigid muscles, the screaming ... oh my! The rage of a toddler quite takes your breath away.

When the tantrum is spent the child sobs, gulps, hiccups then falls asleep all hot and bothered to wake-up relaxed, refreshed and breathing normally.

However, temper tantrums are not socially acceptable so, as children, we learn not to scream, kick or punch just because we can't get our own way.

So we learn to hold onto our breath and our anger and our frustration and as a result our muscles tense and stay tense – including the diaphragm - and we stop letting go.

What happens over time is all that pent-up energy stays locked inside the chest, jaw, neck, arm and leg muscles. Ouch!

EVENTUALLY we begin to live, work, sleep and walk with a lifted rib-cage, clenched teeth and tense shoulders, so instead of our diaphragm breathing for us our pectoral muscles in our upper chest have taken-over.

Now, our Pecs are those large bulging chest muscles that body-builders like to develop and gorillas thump to show how tough they are.

Yep, Pecs are the muscles used for swinging the arms – for upper-body strength, fighting and punching. They are our angry muscles!

They are also connected via other muscles to our upper rib-cage and can lift and expand the rib cage thus mimicking the diaphragm but there's an important difference:

Breathing with our chests only expands the top third of our rib-cage so our lungs only partially fill with oxygen.

Added to this, our lungs are bell-shaped – wider at the bottom than at the top so chest-breathing only fills the top narrow lung tissue so it's a double whammy.

We even catch ourselves holding our breath in a crisis, as did Phil.

This is bad news because it is at those moments of red-alert that we need all the oxygen we can get.

In other words, our muscles respond to our emotions, we

smile and laugh when we are happy; we scowl when we are anxious, cross or confused.

Get the picture?

The thing is this ... our emotions today are exactly the same as our cave-dwelling ancestors, It doesn't matter if it's a failed hunt or not having enough money to pay the bills, we become anxious.

The point is this, these 'causes and effects' of fear and anxiety are hard-wired into our biological-emotional loops.

To be honest, without these loops we wouldn't have survived down the ages and very often it's the knee-jerk reaction that saves us from disaster – leaping out of the path of a bus, freezing when stalked by a sabre-toothed tiger.

HOWEVER, humans frequently make knee-jerk reactions to a situation that, in turn, create negative emotions in others - like fear and anger, so our behaviour becomes cycles of destruction.

Recognising these cycles is one thing, changing how we react to our fear and anger is another and it's not easy to change, but it's not impossible.

A little self-knowledge and awareness makes all the difference ...

Ever watched a scary movie and gasped with fright, spilt the popcorn and felt your heart race? Ever noticed how you hold your breath until the 'moment' has passed?

That's the ancient biological fear response kicking-in and yet it's only a *film,* it's only make-believe, for goodness sake!

Fortunately, once the film has ended we can relax and talk about it and say we enjoyed the experience and that's because the adrenalin rush has passed and our bio/emotional loop now informs every cell in our body that we're *safe.*

Not that we're aware of this – we're too much in our heads

to be conscious of what our bio-emotional-chemical automatic response is up to.

Now, you're probably thinking – 'what has all this got to do with Homo spiritus? If breathing and the adrenalin-rush is automatic and unconscious what's the big deal?'

We'll come to that shortly – it is an astonishing 'big deal', nothing short of magic and it is a clue to our uniqueness; but baby steps, right?

Here's a hint - what happens between those two momentous occasions of our first and our last breath is up to fate and the *choices* we make.

When Phil was alarmed at hearing an unexpected noise he held his breath – his diaphragm 'froze', it too was 'listening' and waiting for what would happen next.

Why?

Because what happened next would depended on what Phil *chose* to do. He could have run out of the building; he could have phoned the police; He could have told his dad; he could have called out.

Instead, he chose to *investigate* and what happened after that depended on what he found behind the closed door so he continued to hold his breath.

It's like his whole body was in suspense, which it was. When we faint with fright this the body's extreme reaction to this kind of situation. The autonomic nervous system takes over and down we crash – no choice there.

Phil, however, was confident of his strength and ability, plus he was a young male adult so powerful hormones flooded into his blood, preparing him for 'fight'.

If it had been burglars, well, that would have been a different story and Phil may not have come out of it too well, unless, of course he was also trained in Martial Arts.

On that note ...

Breathing is the starting-block for many ancient and modern health practices.

In China, the universe is viewed as a system that functions as a flow of opposites called Yin and Yang and if you think about it, breathing-in is the opposite of breathing-out.

So, for example Yin is anything that is pale, cold, wet, dark, the moon, blue, down, in, slow, passive, still, receptive, female, breathing-in.

So, the opposite Yang, is bright, hot, dry, light, the sun, red, up, out, fast, active, motion, dynamic, male, breathing-out.

Those of us who have tried yoga, T'ai Chi or meditation, or a martial art such as kung fu or jujitsu know that they all work with the breath.

Try this experiment:

Stand facing a wall, some feet away and then lean forwards with the palms of your hands flat on the wall and your arms straight.

Now take a deep breath-in and at the same time apply all your strength into the wall as though to push it away from you.

Keep up the pressure and then breathe out and notice the difference – you can apply a greater pressure when you exhale – Yang energy.

This is why martial arts use the out-breath as a tool of strength often accompanied with a shout!

It is impossible to shout on the in-breath, it is impossible to talk, sing or whistle effectively on the in-breath.

Teachers are often oxygen starved at the end of a day – all that talking, all that breathing-out, all that Yang activity and not enough Yin – not enough oxygen. Headaches, stiff necks and exhaustion become the norm.

A period of Yin – of quiet and deep slow breathing – a walk in the park or the beach or to sit somewhere peaceful for twenty minutes 'zoning-out' can quickly and efficiently restore the balance – the harmony.

So, you see, breathing can be seen as a flow of opposites, we breathe in (Yin) and we breathe out (Yang). Rather like the tide that ebbs and flows, the moon that waxes and wanes.

Our tummies should rise and fall as we breathe; the chambers of the heart empty and fill with blood and all these natural rhythms work fine until the balance is disturbed and becomes either too Yang or too Yin.

High blood pressure, panic attacks, insomnia, acid-reflux, inflammation, a high temperature and a racing pulse, aggression and excitability can all be seen as Yang symptoms.

Lethargy, chronic fatigue, low blood pressure, constipation, under-active thyroid, depression, fainting, anaemia, temerity, can all be seen as Yin symptoms.

When we suffer from these symptoms we feel stuck with them. Actually, it begins as a slow and steady habit, happening almost without our awareness until BAM!

We have a crisis and end up at the doctors or in hospital and often prescribed medication to help restore the body's homoeostasis – a biological term for balance.

It can seem as though the rest of our lives are doomed to a certain level of un-wellness and medication, like living in a cage.

But it needn't be.

By understanding the importance of correct breathing, recognising when it becomes dysfunctional then knowing HOW to correct it can be the key that unlocks that cage.

A key to better health, more energy, good digestion, refreshing sleep and ultimately to another door that unlocks an even deeper mystery.

Sounds wonderful?

It is!

So, here's what we do:

We need to re-learn how to breathe

Now, try this second experiment:

Place one hand on your chest and the other under your rib-cage and wait to see which hand is moving as you breathe. Try it now.

What do you mean nothing is happening!? BREATHE for goodness sake, just breathe ... go on – take some bigger breaths, nothing huge.

Now did you feel something happening? Was it your upper chest, your upper rib-cage that lifted? Yes? Well at least you were taking in some air but not nearly enough plus all the wrong muscles were working.

Try again, only this time, push your tummy muscles in and out – this will wake-up your diaphragm.

Now, as you breathe-in push your tummy muscles out – encouraging your lungs to fill-up. Your upper hand, the one resting on your chest, should be still, so make sure this area is not moving.

Now, breath-out and push your tummy muscles in towards your spine, this will squeeze the air from your lungs. Once again, your hand resting on your chest must not move.

This method of breathing will feel forced and unnatural but stick with it.

Occasionally take an extra deep breath and this time your chest will move as the upper lungs inflate and deflate as you treat your body to an extra dose of oxygen. Fantastic!

What this 'tummy-breathing' exercise is doing is teaching your diaphragm how to work again.

It is how we should be breathing when we are at rest or just sitting not doing anything – like stuck in a traffic jam, watching T.V. As with any form of exercise it takes concentration and practise until it becomes second nature, and it will ... but not over-night.

Most importantly, during the day start to take note of when you're *not* breathing – usually when we are multi-tasking or late or anxious like stuck in that traffic jam ...and ... BREATHE.

Drop your shoulders, relax your arms, shake your wrists, unclench your teeth and BREATHE – from your tummy.

WHY must we breathe from our tummy? Because the lungs are bell-shaped, remember?

Also, the right lung is larger than the left, not that you need to know that ... it's because the heart chamber takes up room on the left side of the chest cavity, you didn't need to know that either.

The point is this, being wider at the base the lungs need to fill-up with air from the base upwards and we need to be using our diaphragms to achieve this.

When we breathe inefficiently - using our upper rib-cage and Pecs we only fill the top narrow part of our lung – if we're lucky - and that's not sufficient air in-take and so we become oxygen starved.

Everybody's heard of aerobic exercise which is all about working our heart and lungs extra hard by doing vigorous work-outs. This is when we DO breathe with our upper chest and Pecs as well as the diaphragm.

During vigorous exercise our leg and arm muscles quickly exhaust their supply of oxygen and start to fill-up with lactic acid so some emergency breathing is required to get an extra supply of oxygen and eliminate those nasty waste products.

We pant and gasp, get stitch, cough-up phlegm and feel like we could *die*.

Unless, of course we are used to aerobic work-outs then it isn't nearly so painful and actually gets to be quite enjoyable ... even *very* enjoyable though it's best not to expect too much too soon.

However, to do some vigorous exercise two or three times a week is a pretty good idea if we spend our life sitting in front of computers, in cars, coffee bars and watching the telly (pretty much of all us, I guess.)

And that's not all ... wait for it ...

There are two wonderful by-products of correct breathing

One by-product of all that vigorous exercise is that the lungs get a spring-clean. Why? Because a yucky pool of mucous, dust, old bacteria, micro-particles from pollutants such as hair-spray, noxious fumes and other chemicals all settle at the base of our lungs.

All this muck gets shifted when we exercise – hence the hacking coughs when we first start our new get-fit regime!

The good news is that the cough; the stitch; the red-face and jelly-legs really do go away – gradually – after a few weeks. The heart and circulation quickly improve their performance along with the diaphragm and other breathing muscles.

Here's another little known fact - correct breathing also

improves our digestion! Whether after exercise or just everyday breathing we can begin to see a difference in how our gut functions.

If you've taken on-board that stuff about Yin and Yang you'll gather that aerobic exercise is a Yang activity, unlike Meditation, which is a Yin activity ...

... both hugely beneficial.

SO, if you suffer from something like constipation (Yin) then doing something Yang is going to help redress the imbalance. Exercise makes us warm, we jiggle about, it gets things MOVING!

So what about the Yang (inflammation) digestion problems such as acid stomach re-flux, peptic ulcers, colitis, I.B.S?

Now, here's the secret. When we breathe with our diaphragm, this wonderful muscle very kindly massages our digestive organs lying beneath it.

The diaphragm, remember, is the ceiling of our abdominal cavity and when it rises and falls it is gently pushing down on our stomach, liver, pancreas etc.

This gentle movement encourage these organs to secrete their enzymes, juices and hormones, as well as increasing their blood flow in a nice warm and relaxed fashion.

SO, if you suffer from any of the above Yang symptoms then the kindest thing you can do for yourself is to sit or lie down somewhere warm and comfortable.

Place you hands as before, one on your chest and the other on your tummy, and focus on breathing with your tummy muscles – remember this will activate the diaphragm.

Don't be surprised if you take a sudden deep sigh right up into your upper chest cavity.

If this happens it is an excellent clue that your body is starting to regain its balance - its homoeostasis, all by itself and this is happening because you are *allowing* it to happen.

Any time of the day, anywhere, you can chose to breathe correctly and when you do, over time, your body can start to heal itself of all its little niggles.

One of the best times is as you settle down to sleep on a night and also just before you get out of bed in the morning. This will remind the diaphragm to work correctly as you sleep and at the start of each day.

After a few minutes of this tummy breathing don't be surprised if you start to hear lovely gurgling noises, it means everything is WORKING! The stomach, the liver, the bowls ...

So, it makes sense to be an excellent breathing machine.

AND it doesn't end there because we are, of course, much more than a machine responding to internal and external influences.

We have free will, remember.

We can make choices.

We can, if we try, choose to breathe correctly.

We can, if we practice, unlearn our dysfunctional breathing.

WE, Homo sapiens can CONTROL how we breathe and becoming aware of this is the first step to our continued evolution.

Why?

Because there is another extraordinary fact that makes humans unique – here it comes, it's a biggy!

Take Control of your BREATHING and you take control of your *LIFE*

Remember, we humans have will-power, we are also CONSCIOUS of breathing, we are CONSCIOUS of our existence, we have a strong idea of Self.

If we didn't have a strong idea of Self we'd be like our

cousins the chimpanzees and gorillas and we wouldn't whistle, sing, laugh or cry.

All these human traits happen because of the connection between our personalities, our emotions and our muscles; our breathing and our will-power. It is extraordinary and a key to our health, happiness, personal fulfilment and ultimately the dawn of Homo Spiritus.

(It's seriously worth reading that last paragraph again.)

Okay, so we are emotional creatures with will-power and are constant victims of our own emotions.

And if that isn't tough enough we are also victims of other people's will-power and emotions – that's for sure! And vice-versa, don't forget.

We often have knee-jerk reactions and make a snap decision born out of anger, fear or resentment which we may later regret.

Now then;

An extraordinary thing starts to happen once we learn to breathe correctly. We find we begin to control our lives in a much more powerful and meaningful way.

It starts like this: as we consciously become aware of our breathing we start to realise *when* and *why* we are not breathing correctly and this makes us examine not just the situation but also our emotional response.

Picture the scene...

It is eight thirty in the morning; a suburban kitchen-breakfast room is in chaos because nineteen-year-old Ellie has lost her purse.

The back door is open and Ellie's Mum is sitting in the car, on the drive, the engine running and her own patience running out. She toots the horn.

Ellie yells from the kitchen. 'For god's sake!'

She throws the cushions off an old sofa. 'Give me a minute!!'

*Her Mum winds down the car window. 'Ellie, I'm going in ten seconds or we'll **both** be late for work.'*

'Wait!!!'

Her Mum sighs deeply, biting her lip; her daughter is always losing things.

'Ellie, we've searched everywhere. I have to go. Are you coming?'

Ellie swears.

Her mother lifts her foot off the clutch and drives away.

Ellie begins to sob with rage and frustration, she slams shut the back door, swings round and her elbow catches a bowl of sugar near the kettle, it tips over and spills sugar everywhere.

You've never heard such bad language.

The back door opens gently and a head peeps round.

Ellie doesn't notice, she's collapsed on a chair and covered her face with her hands.

A woman with purple hair walks in, thirty-something, dressed in leggings, T-shirt and a purple tabard; a pair of rubber gloves in her hand.

She takes in the chaos. 'Ellie?'

Ellie looks up. 'Suzi,' she wails.' I've lost my purse.' Jumping up she starts pacing, her face red, flinging her arms in the air. 'Mum wouldn't wait and my boss is going to kill me and ...

I'll help you,' says Suzi, calmly, and taking Ellie by the arm, sits her down.

Ellie starts pleading with her. 'Will you clean up the sugar, Suzi, please? It's all over the floor and ... and ...

Ellie starts to hyper-ventilate.

'Yes, but first do something for me. Look at me, Ellie'. Says Suzi.

Ellie looks into the calm blue eyes.

'First you've gotta start breathing properly.'

'What?'

'You're oxygen starved, you brain can't think, your body's toxic with adrenalin.'

Ellie's eyes widen.

Suzi pulls out a chair and sitting opposite Ellie shows her how to relax and to breathe correctly and to drop her shoulders and release her jaw muscles.

It's a struggle but Suzi gently persists until Ellie is more calm and relaxed.

Finally Suzi smiles. 'Now then, what is the most important thing to do right now? Think calmly.'

Ellie takes a deep breath and thinks. 'Um ... I ought to let my boss know I'm going to be late.'

'Okay, so let's phone her.'

'My mobile phone is in my purse!' Ellie's voice starts to wail again. 'And my keys, and my credit cards.'

'Hush, relax and breathe. Okay, send her an e mail. Here.'

Suzi takes her I-phone out of her pocket and switches it on.

Ellie takes it. 'What do I say?'

'That you're going to be late, you've lost your purse, missed your lift and you'll work your lunch hour, that you're very sorry and it won't happen again.'

Ellie scowls. 'You don't know her, she's a dragon. She breathes fire.'

'I'm sure she's doing the best she can.'

Ellie looks at her, puzzled.

'It's what we all do, until we know how to do things

better. So, when did you last have your purse?'

'Um ... well, I had it when I came home from work yesterday. I got home before Mum and let myself in.'

'And later?'

'Nothing, I didn't go out last night. Wait! Yes, I did ... Mum and I fancied some chocolate-chip ice cream ...'

A minute later Ellie finds her purse in the bottom of a scrunched-up carrier bag. Two minutes later a taxi is ordered and Ellie is sweeping up the sugar. She watches Suzi at the sink.

'Suzi, your kinda smart, why do you clean people's houses?'

'Because I like working for myself and I discovered the joy of cleaning.'

'How come you're so ... sorted?'

Suzi laughs. 'I didn't used to be, I used to be a right mess. Then one day I met an angel.'

Ellie's eyes popped.

A car horn tooted.

'There's your taxi, Ellie.' Suzi takes a small card from her pocket.

'And take this - I hold stress-busting classes on a Thursday evening. You'll be every welcome'.

So, next time we become aware we're not breathing correctly ... maybe we've chosen the 'slow' check-out at the supermarket; the car's running on empty; someone's 'got up our nose' or hurt our feelings.

Maybe we've sucked in a lot of air because we're about to yell at someone ...

STOP - we need to drop our shoulders, unclench our teeth and fists: stop frowning and force our diaphragm to work. Breathe deeply from our tummy, CONSCIOUSLY keep breathing nice and deep and slow and every time we breathe-out we let our muscles relax that little bit more.

The fact that we're breathing calmly and the muscles in our arms, shoulders and jaw are relaxed tells very cell of our body - 'hey, guys, no problem here, we're cool, we're calm, switch off the red-alert'.

And that, my darlings, is exactly what starts to happen – the loop is reversed!

Blood pressure lowers, hormones switch-off and suddenly we feel calmer and in control because we *are* in control.

Okay, the situation hasn't changed but our response to it has:

The check-out queue is no big deal – we're not here for life.

This isn't the first time we've run the car on empty - maybe it's time to keep an eye on the fuel gauge.

Hurtful words are just someone else's knee-jerk reaction - It's not personal - we can handle this.

Yelling and getting angry won't help the situation.

So, not only do we deal with the situation better but we're safe-guarding our physical, mental and emotional health (and possibly others, too) from all that mega-high ancestral stress response.

Remember we have a choice, we can have a knee-jerk reaction to a situation and get a stress-head like Ellie, OR we can practice nice tummy-breathing, let go of tense muscles and stay calm.

We get to choose.

This new approach to the stress in our lives doesn't happen over-night, but it is fun to practice, trust me.

Here's an exercise:

Next time you watch a scary or nail-biting film, check how you emotions kick-in as you get caught-up in the action and how your body is responding.

Yep! You're all bunched-up, on the edge of your seat, you've stopped breathing, eyes are as wide as saucers ... so, force yourself to lean back and breathe from your belly, just to see if you *can* switch off your body's stress response.

Or maybe you witness a scary situation that is for real – like drunken football supporters or a wild hen-party getting on your last train home – aargh! You immediately hold your breath and feel your stomach twist.

The same thing applies; drop your shoulders, breathe and take stock before reacting to that knee-jerk response then calmly move away from the situation if you wish, the choice is yours.

That's the magic of being human, you can choose.

You can choose to learn how to breathe correctly, how to command our diaphragm and how to relax the muscles in our neck, jaw, shoulders etc. and when we do ... everything else follows.

Also, have fun practising belly-breathing when settling down to sleep. Remember to breathe when feeling anxious, busy or stressed because that's precisely when the brain needs *lots* of oxygen, not *less*.

Become a super-breather!

One last and incredibly important thing ...

HOMO-SPIRITUS – ARISE!

As we start to reverse this breathing loop of our ancestral response to negative emotions, our family, friends, work colleagues, strangers ... anyone who comes in contact with us, will sense we're 'different' and be impressed by our calmness and control.

This super-cool control doesn't mean we're going to start behaving like an emotionless Zombie.

Nothing could be further from the truth, and of course it doesn't remove the source of the stress, just its harmful effects.

We become the stuff of heroes – like Suzi. Think of all our favourite film heroes; do they get stress-heads? No way.

Okay, so their roles are being played by actors and actresses, but that's just the point, at this stage in the rise of Homo-spiritus we are forcing ourselves to remain calm.

We're 'acting' in a way that is not our normal response to life's hiccups and if we just add a touch of humour and a hint of humility then ... wow! We will really start to tap into something very powerful indeed.

Another tremendous advantage to taking back control of our lives is we begin to *believe* in ourselves which is the first step towards fulfilling our true potential because there is much more to being human than simply controlling how we breathe.

Meanwhile we shouldn't beat ourselves up if our breathing goes to pot and we occasionally start running around like a headless chicken.

Eventually we will find breath-control becomes second nature; we *change*.

As we go through life, make friendships, have partners,

perhaps raise children or grandchildren, we will be able to share with them this trick of breath-control and thus strengthen its imprint at a quantum level – our human genome.

That is what evolution is all about and happily there are other bio-emotional loops programmed into our behaviour that we can consciously reverse.

Excellent news!

We are now ready for Step 2.

STEP 2

STRESS LESS

Stress has a huge impact on our lives so let's take a closer look at what and where it comes from and why it can wreck our lives

The point is this – stress is as natural to life as the exchange of gases when breathing.

Simply surviving the forces of nature and finding enough food and somewhere safe to live and grow is a very stressful business for all creatures.

All life is programmed to survive stress and it doesn't matter whether you're a seed, bacteria, fish, an insect or a mammal.

Mammals however have an emotional response to stress as well as a biological response and, as we have already seen, things can then get very complicated.

When it comes to Homo sapiens stress can become a NIGHTMARE!

Understanding a few basic facts about stress can help us adopt a new approach and by reversing its bio-emotional loop life can become immeasurably easier and ultimately joyful.

Firstly, stress is all about survival.

Look at it this way, Life is a tough ball-game. Our planet home is hurtling through space at around 60,000 mph, spinning on its axis, with a molten core, an unstable crust, extreme weather and a strong force of gravity.

Add to the fact that soft-bodied animals (us included) can easily and quickly bleed to death and depend on a constant supply of atmospheric oxygen to stay alive – it's a wonder that any reach adult-hood, let alone a ripe old-age.

On top of all that animals need to EAT! Obtaining enough to eat has been a constant stress factor.

Also, it seems that living-things have tendency to eat or consume other living- things, which means all life, other than plants, have to run, hide or fight so as not to get eaten - MORE stress!

Another life-essential is plentiful supply of water which isn't contaminated by animal waste, harmful microbes or parasites.

THEN, we factor in all the other bacteria, poisons and viruses that can, and do, invade our warm, fluid (65% water) bodies and we wonder how we survived evolution at all.

Homo sapiens also have a long history of killing one another.

Yep, it really is a marvel we have survived this long (so far) and the only reason we have, (since we lack sharp teeth, claws, stings, fur, strength, speed and flight) is because we developed a thinking brain and learned how to combat most of our nightmares.

We learned, for example, how to make shelter for protection from the weather; tools to hunt with and clothes to keep us warm. We learned how to dig wells for water and to how make fire to cook meat and to experiment with herbs to aid sickness and injury.

So, apart from the occasional earthquake, ice-age, flood, famine and drought, life should have become easier and less stressful for our clever nomadic ancestors and their stone-age descendants, followed by *their* bronze-age descendants and *their* iron-age descendants.

Most of who gradually became less nomadic and started to farm animals and grow crops so that finding food became easier and safer.

And it didn't stop there!

With the development of science and the discovery of gases, other metals and minerals we progressed towards the Age of Enlightenment and then the Industrial Age with its huge leaps in machinery, science and medicine.

And now we find ourselves in an Electronic Age, rapidly becoming the Quantum Age, with all manner of Artificial Intelligence and robots, an Age where nothing seems impossible and a future of space travel to new worlds and even' immortality' doesn't seem too far-fetched.

So, are we less stressed?

Are you kidding!

More and more people are being medically treated for stress-related illnesses, both physical and mental and there are lots of reasons for this but first we need to talk about SEX.

Life is all about sex

Why do we need to talk about sex?

It's because sex links humans to all life and helps us understand our emotions and until we understand where we are coming from we've no hope of dealing with where we are going.

There's simply no escaping the fact that Life is hard-wired to produce more of its own kind.

Yep, Life is all about sex and since there is a strong risk of getting eaten or being killed due to storms, floods, drought, famine, earthquakes, hurricanes, injury or disease there is enormous pressure to reproduce as much as possible as quickly as possible.

And then there is the stress of parental care which involves feeding and protecting your offspring - more stress and for some species this can last weeks, months, years and with humans ... forever!

However, there is another more basic reason why sex is so stressful.

Sex = Competition = Stress.

We see it everywhere, don't we? All animals compete for a mate, territory and food.

You only have to see two male polar- bears fighting it out, all bloodied on an ice-floe, to realise what's at stake.

Or two magnificent stags that are the best of buddies most of the year and then try and kill one another during the mating season.

Hmm!

It would seem there is never enough to go around – not enough land, not enough food, not enough water, not enough males or females.

Hmm!

So, if all the things that most species, including humans, need in order to be safe, comfortable and sexually satisfied are in short supply then we're going to have to compete for what we want and since we all want the same thing it looks like we're going to have to work very hard to get it.

Or fight for it.

Then there is the stress of protecting what we possess so it isn't lost, damaged or stolen. In fact, competition is never an end-game for humans because being so clever and inventive, far-seeing and calculating we can anticipate trouble and thus prepare ourselves for the worst-case scenario, be it weather or war.

The weapons we've invented as part of this 'gain and protect' scenario is mind-blowing - from spears, axes and swords to canons, bullets, bombs, nuclear and chemical warfare.

And the result of all this is ...

We create hell and even those who do survive the war-zone are left with destroyed lives, homes, cities, crops and livestock = more scarcity and so it starts all over again. Loop-de-loop-de-loop.

Competition is stressful in other ways, for example we tend to think that to be beautiful, successful, rich and famous will make our lives easier.

That may be true but ask any celebrity and they will tell you they know all about stress!

There is something else that is unique to Homo sapiens, we often don't know when sufficient is enough and enough is sufficient.

This can make us greedy so maybe we need to ask ourselves if greed, due to fear, has also become encoded into our behaviour.

A Greed gene will make us want more and more which is why Competition becomes a stress-ridden anxiety even during times of peace and plenty.

Can our modern lives ever become less greedy, less stressful and more secure, peaceful and fulfilled for the individual and society as a whole?

The answer to that is 'yes' though we have to chose to want this and then re-program ourselves.

How?

Read-on.

The stress response needs to switch-off

The biological stress response was never intended to be constant.

Stress should not be an all-day, every-day event, it should peak when an emergency occurs and then switch-off when the emergency has passed.

Have you ever watched a film of a lioness hunting a herd of grazing zebra?

The herd is put to flight as the chase begins. If the hunt is successful the lioness will take-down a young or weaker Zebra by gripping its throat in her strong jaws and the zebra dies quickly.

What is so astonishing is the reaction of the remaining herd. They stop panicking as soon as the lioness has stopped chasing them, twitch their ears then continue grazing as though nothing has happened ...

... 'Hey! Your buddy has just got killed and is being eaten! Where's the trauma, the grief, the sadness?'

But of course a Zebra is not human. For them the stress is over and life goes on, as nature intended.

Not so for humans.

Remember, we have evolved into emotional beings with a strong sense of Self and of the future, hence we suffer.

We should be the most relaxed and laid-back creatures on earth.

We are at the top of the food-chain.

We can predict the weather; build safe homes, towns and cities.

We can find cures for diseases and antibiotics for germs, antidotes for poisons and ways to keep warm in the winter and cool in the summer.

Water is made chemically safe and delivered into our

homes and farming is so scientific there is no reason why anyone should go hungry.

Modern medicines mean we can mend broken bones, replace organs, attach artificial limbs, correct poor sight, hearing, dodgy circulation and digestion and we can even perform genetic engineering.

Advances in psychology and psychiatry delivers numerous Counselling therapies to help us understand ourselves and give support at times of mental and emotional crisis.

Many of us now live well into old age and travelling is both quick and relatively safe.

On top of all that we are supposed to have more leisure time than ever, due to labour-saving gadgets and machines, not to mention robotics and computers and thus able to enjoy a huge variety of sporting activities, holidays and entertainment.

And yet ... and yet ....

Are we truly happy, content and fulfilled?

Do we ever have a time with no stress at all?

It would seem not. The truth is we worry constantly; in fact we have evolved to worry and for good reason, we all know what happens if we walk around with our heads in the clouds - we walk head-on into trouble.

Worrying is about trying to anticipate and avoid a crisis, so, in this modern age of advanced science and technology what is that we worry about?

Well, here is a list of the kind of things humans worry about:

'Acts of God' – floods, fire, earthquakes etc
Accidents
Ageing

Change
Crime
Criticism
Debt
Disease
Failure
Family crisis
Governments
Growing old
Heart-ache
Hunger
Illness
Loneliness
Loss
Mental illness
Not being in control
Not being loved
Our loved ones
Pain
Politics
Poverty
Reputation
Rising Interest rates
Safety
The unknown
War

You won't find a zebra worrying about such things, and no doubt you can think of other things. The thing is we recognise all these 'worries' because they are familiar to us.

Children too have their own worries and quickly learn patterns of behaviour from their parents, siblings and other family members.

If a child has a stressful and unhappy home-life, then

their future can look bleak.

If a child has a calm and contented home-life, then their future can look hopeful.

School-life, of course, brings more worries as competition; fear of failure and rejection are reinforced year by year. Schools have massive social problems with stress and anxiety high on the list for both staff and pupils.

All the techniques described in this book can help schools enormously. The younger a person learns how to reverse the loops of our bio-emotional behaviour the happier society can become.

Also, the more we gain in life the more we have to lose, so we worry about that, too and it's not just our possessions; reputations and jobs too can be lost in an instant.

We have too much debt ... we start feeling out-of-control ... and once the anxieties start piling-up and with little hope of change - we're in trouble.

Sometimes these can turn into long-time situations, even life-time situations and cause crippling health problems. Such as:

Accident-prone
Acid reflux
Back ache
Constipation
Depression
Diarrhoea
Fog-brain
Food / drink /narcotics cravings
Headaches
High blood pressure
Indecision
Indigestion

Insomnia
Irritability/anger
Lack of confidence/self esteem
Lowered immune system
Night sweats
Poor concentration
Poor judgement
Tight shoulders and neck muscles

Oh boy!

Hands-up anyone who has ever suffered any of the above? Hands-up anyone who is suffering some of the above right now?
Hmm.
So, whether it's unpaid bills, or the habit of *always* being late, or a bad-hair day; exams looming or someone we love has been in a car-crash or we dread going to work … to school … going home …
… Or maybe it's a long term situation like caring for a disabled child or parent; an unhappy relationship; redundancy, prison, house-repossession, bankruptcy …
… Or maybe it's something a little less rational, such as social anxiety or a phobia or an obsessive compulsive disorder.
The point is, our bio-emotional response is always the same – hormones such as adrenalin and cortisol pour into our blood stream, acid is released into our stomach, blood-sugars surge and our blood pressure rises.
Unless all this is dealt with we may experience not only physical illness but possibly a mental/nervous breakdown, which will need long-term medication and extremely understanding family and work colleagues.
So, what's the solution?

Hope you wake up one morning and find the stress-factors have magically disappeared?

Nu huh, life isn't like that.

We can *pretend* we haven't maxed out our credit cards.

We can *ignore* the fact that the house is a tip.

We can leave mail *un-opened*.

We can *kid* ourselves we're in a healthy relationship.

We can drink coffee, eat chocolate, smoke another cigarette; buy another pair of shoes; keep washing our hands ...

We can do all these things to reduce the bio-emotional stress response – a quick blast of endorphins may give us a brief lift but then comes the crash and so we need more coffee, more chocolate, more cigarettes etc and before we know it we're drowning in debt, shoes, clutter and run the risk of severe depression, obesity, lung cancer ...

Whoa!

STOP!

Let's take a deep breath, from the tummy and right up into the chest, breathe out and pause a minute.

Okay. So we're realistic – we know the causes of all our stress are not going disappear with a wave of a wand so what are we going to do about it? Curl-up in a ball and cry?

Crying does help, it's a way to let go – lots of diaphragm movement and shoulder-shaking that can release the tension - but it's not a solution to how we feel; crying is not the best way to reverse the harm that the stress-loop does to us.

So, how do we do it? Well, correct breathing is a massive step in the right direction and may prevent us from acquiring *more* stressful situations. (Remember the cool, new, clear-thinking person you are about to evolve into?)

Remember how tummy-breathing 'tricks' our bio-emotional feed-back so we don't kick-off, over-react or panic?

Good, because this next 'trick' is just as powerful; a little more mysterious and a lot more subtle and takes us one more step towards the rise of Homo spiritus

Once again it's all about certain muscles and our ability to choose.

Smile and the world smiles with you

(*Smile though your heart is breaking...* sung by Nat King Cole, Rod Stewart & others)

You'll remember we talked about the diaphragm and how humans cry and laugh and sing and whistle? Well, the muscles in our face help us do these things.

The musculature in the human face is remarkable!

Our face reflects all our emotions.

With just the tiniest of changes in our facial muscles we can express boredom, joy, amusement, sarcasm, irritation, anger, frustration, surprise, shock etc.

What is equally amazing we can control these muscles by choice!

Try lifting one eye-brow, it speaks volumes!

Think of actors – they can show by their faces how the character is feeling. They PRETEND to be angry, scared, amused etc and are so convincing, we, the audience gets caught up in the actions and emotions as though it is all real.

So, the next step to reverse the stress loop is:

To smile

We now understand that our body and emotions are intimately linked, they 'listen' to one another all the time. It is a sub-conscious, autonomic loop.

Our emotions are reflected in our muscles - we feel angry so the muscles in our face form a scowl and our lips tighten. We feel bored - we yawn. We feel tense - we stop breathing. We feel surprised - our eyes widen. We feel shocked – our mouth drops open.

The reverse is true – if we have a headache we feel miserable, if we fall and hurt ourselves we feel shocked and tearful.

Physical pain triggers an emotional response.

Emotional pain triggers a physical response.

So what happens when we feel a pleasant emotion - humour, tenderness, pleasure? Well, our muscles soften, we relax and maybe we smile. The stress hormones switch off and the feel-good factor kicks in making the world seem a better place.

Again, the same is true for physical pleasure – basking in the warmth of the sun or having a hug can trigger those pleasant emotions.

Physical pleasure triggers a happy emotional response

A happy emotion triggers a relaxed physical response.

Because Humans have developed such a strong sense of Self we cannot separate the mind from the body.

End of.

But don't stop here, read on:

We now need to understand the extraordinary power of a smile

Smiling is a reflection of our positive emotions and is unique to humans and this is worth thinking about.

No other animals have facial muscles quite like ours or the ability to control their breathing; or the ability to make life-changing choices; or such a strong sense of their own existence or the concept of time.

Here's a simple exercise to do, right here, right now:

Frown a deep scowl ... imagine you're really cross, frustrated or annoyed and feel how the muscles in your face and around your mouth change and also your eyes!

Do your eyes seem a little hard, mirroring a subtle but significant change in your perception of the world?

Now, continue to hold that angry face and try to smile. Difficult, isn't it? In fact, the muscles around your mouth struggle to make that smile. To be honest it isn't a smile at all, it's some kind of ghastly grimace!

Now, slowly let go of the 'anger' - the scowl and pursed lips but continue to 'smile. As your brow begins to 'clear' and you let go of the anger your smile will become much more genuine.

Now ...hold that smile, your beautiful smile and feel how your eyes soften.

Feel how your shoulders drop, feel a small but very subtle change in *how* you feel. You may feel like taking a deep breath – that's good!

Take another breath and smile a little wider – you have just given yourself a dose of endorphins without eating chocolate!

You see, a smile is a signal that 'all is well'.

A smile is so powerful that it sends a subtle message to every cell in your body to switch off the stress response and it doesn't have to be a big cheesy kind of smile.

Trust me, the tiniest smile, a Mona-Lisa smile or less, is all it takes. Even smiling to your self can work magic. Even having the Intention to smile can work wonders!

A smile 'tricks' your body that you're 'happy'.

This is the secret of reversing bio-emotional loops. The causes of our stressful life still exist – we just trick our *body* into believing life is not stressing us.

Sure, smiling when stressed is going to feel like a lie – we dread opening the gas bill – so, we take some deep belly-breaths, unclench our teeth and now we try and smile.

Nah, that doesn't feel like a smile, this is not a happy experience.

Stick with it – remember, we're not trying to fix the bill (that's for later, in Part 2).

We're trying to stop stress from killing us.

Yes, in a way we're acting and it's not easy but in time as you practise this trick it will become second nature.

Remember:

A smile reverses the loop of a negative stress reaction.

And then what happens?

We Change

One of my clients is a singing and music teacher and she taught this loop reversal to her choirs and their singing became transformed. Smiling not only helped negate their nerves when performing to the public but also improved their performance and enjoyment of the occasion.

When we have troubled thoughts (day or night) then remember to smile that tiny smile and those pesky thoughts just vanish! Sure, they creep back again and so we smile again.

The diaphragm also relaxes when we smile (it thinks we're happy, right?) so we may find we take a sudden deep breath at the same time and that's wonderful news. That rush of oxygen to the brain can help us to look at and resolve our problems more efficiently.

As we continually practice letting go of our bio-emotional stress-reaction by breathing and smiling we begin to change how we view life and ourselves.

By *consciously* responding to stress by reversing these loops of dysfunctional breathing and scowling we can control our lives so much better. And guess what?

It empowers us.

We become gentler in our attitude towards ourselves, our life and towards others.

We become more patient, more understanding and then something very powerful starts to happen and it's all down to Body Language.

Body language says it all

Body-language is a very powerful communicator, more so than words. This is especially true of humans. We can assess someone's mood astonishingly quickly just by a small change in their facial expressions and musculature.

Sure, other animals also rely on body language to assess a situation, ever seen two dogs meeting one another for the first time?

Animals however, are also using their extraordinary sense of smell which pick-up powerful hormones that we are no longer capable of sensing.

Our evolution towards a super-intelligent 'animal' bred-out this acute ability to smell these air-born molecules because other means of communication took over.

The reason why humans, particularly males, shake hands on meeting is ancient body language that tells both males 'look my right hand is open; it is not held as a fist; I don't carry a weapon; I am happy to greet you in friendship and to make contact that is not aggressive.'

Hugging and kissing on meeting-up with friends and family tells the same story 'I am happy to get close to you; I know you won't attack me.'

We all recognise the body-language of someone who is either irritated, cross or in a rage and we, in turn, display body-language as a response.

Two people coming face-to-face in a stressful situation can become a minefield of tension.

Picture the scene...

A bicycle storage at a Uni campus – a sudden down-pour has filled the pot-holes with muddy water. Sam races in on his bike and hits a puddle.

'Oy! Watch it!'

Swinging his leg over the seat Sam jumps off and turns at the sound of a man's voice.

A big guy, wearing a long navy coat, smart trousers, brief case under his arm, stands a few feet away; he is splattered with mud and doesn't look happy.

Sam throws back the hood of his cagoule. 'Err ... sorry, mate, didn't see you.'

'Sorry isn't good enough!' says the man, stepping closer to Sam who is a good foot smaller.

'Well ... It's been raining.' Sam mutters

'Idiot! I know that; what I want to know is what are you going to about this ... you've ruined my clothes!?' He gestures to his dirty trouser-legs and coat.

Sam scowls. 'I said I didn't see you ...

'That's no excuse, you were racing like an idiot, I want your name and address because you are going to pay my dry-cleaning bill.'

'No way,' says Sam. 'It was an accident, life's a bitch, get over it.'

The man turns puce and grabs the handle-bars. 'I want your name.'

'Ger'off my bike,' Sam yells, pulling it away from the man.

The man pulls back. 'Well, I guess there aren't that many ginger haired, short-arses like you on this campus.'

Sam grows red with anger. 'Listen, mate ... you can just sod off ...

'Excuse me,' says a soft voice. 'What's going on here?'

The two of them look up and see an average bloke, middle aged, glasses splattered with rain drops, unclipping bicycle clips from his ankles.

'What's it to do with you?' says the big man.

'Maybe I can help?' The newcomer walks up to him and holds out his hand. 'Martin Jones, student council member.'

The big man releases the handle bars and shakes the hand, reluctantly.

Sam takes back his bike.

The big man growls. 'This idiot rode carelessly and dangerously and ruined my clothes. He refuses to give me his name so maybe YOU can do something about it because I want reparation. '

He takes a step closer to Martin who doesn't move and seems amazingly calm and says nothing.

Sam watches with interest.

'I have a meeting with the Dean this morning', continues the big man. 'And just look at this mess!'

Martin looks down at the man's legs. 'Yes, I do see.'

'So, Mr-council-member, what are you going to do about it?'

It seems to Sam that Martin has an invisible smile on his face; his eyes are soft and his breathing calm, even when the big man prods Martin in the chest with an index finger.

Sam's eyes widen. 'Oh boy,' he thinks.

Martin takes a small step back and looks around then back at the big man and this time Martin does flash him a smile. 'What would you like me to do about it?'

'I want this idiot to pay for the dry cleaning!'

'Hmm, but that's not the real problem is it?'

The big man gawps at him.

'And look, the mud is already drying out, may I?' Reaching down Martin brushes the mud off the man's clothes. 'Yes, a stiff brush will clean that up nicely.'

Martin stands up. The big man blinks. Sam stares.

Martin continues. 'No, the real problem is this whole area needs re-surfacing; the pot holes are a disgrace. And it needs to be made secure so pedestrians can't take a short cut through from the car park.'

Smiling again he holds out his hand and shakes the big man's hand for the second time. 'Thank you so much for bringing all this to my attention. Enjoy your meeting.'

The big man blinks again but says nothing and starts to move away.

Martin turns to Sam and says loudly. 'And maybe, young man, we need to put a speed limit in this area too.'

Patting Sam's arm Martin gives him a wink and walks away.

If it's a confrontational situation then it is potentially explosive, as we've just see.

So, if we find ourselves in such a situation and know how to reverse those Loops of bio-emotional behaviour and control our body language as well as our spoken language, we may just be able to diffuse the time-bomb, as Martin did.

This is what happens:

As we listen to what the other person is saying (or ranting about) and we consciously drop our shoulders, control our breathing, release our jaw muscles and smile that tiny smile, we begin, in a very subtle way, to control the situation.

Our 'aggressor' will subliminally read our 'friendly' face; our relaxed body-language, in other words, our calm and un-threatening behaviour.

They will 'see' we are neither intimidated nor angered, that in fact we're in control and this can take the 'wind from their sails'.

What happens next is up to the personalities involved and the situation.

The important thing is that we have reduced the harmful effect that kind of stress can have on *our* emotional and physical well-being, as well as hopefully diffusing the situation.

Wow!

And it can work even if we're not face to face.

On the telephone, texting, messaging ... all these forms of communication can be confrontational.

Here's an exercise...

Next time a customer, client, colleague, friend is 'having a go' at you on the phone, start reversing the loop of your natural stress response by tummy-breathing and

'smiling' as you listen to them.

You will begin to feel calmer and more in control of the situation; it won't be easy and you will have to constantly remind yourself to breathe and 'smile', you will no doubt take a few deep breaths as your body adjusts to this reversal.

Try not to interrupt their 'flow' then, when they pause for breath or snap 'are you still there?' which they will – it will be your turn to speak.

A word of caution... be aware that you too can be the 'aggressor', be aware when *you* start to feel emotions such as annoyance, impatience, resentment, jealousy and anger, and notice how your breathing becomes short and rapid, your fists tighten, your mouth presses into a hard, thin line.

When that happens we need to absolutely reverse the loop. Follow exactly the same techniques; breathe, let go and smile a tiny smile.

And guess what?

You're in control again and find you can handle the situation better, communicate better and no-one gets caught up in the ancestral, knee-jerk reaction of whining or hot-headed anger and the fight or flight response.

By putting the brakes on our bio-emotional response we can pause, take stock and review the situation ...

... 'Okay, maybe I won't respond to that hurtful or annoying face-book post; maybe I'll look at it again later or maybe I'll just ignore it completely'.

When we can do this, mountains then become molehills and all that potential negative energy can be turned into positive energy so we can do all the good things we want to do with our lives.

One last word – we will slip-up, it's only natural but even if we do it's still worth reversing the loop *after* the event.

Here's a true example:

Just the other day someone took 'my' parking space at the supermarket. I had definitely indicated I was turning but this other driver zipped-in and pretended I was invisible.

Was I happy?
Was I smiling?
I was fuming and beeped my horn and continued to fume as I parked elsewhere and stomped out of my car.
Then I recognised my emotional reaction ... so, I reversed the loop.
It took a few minutes but by the time I walked into the supermarket I was calm, in control and saved from an overflow of anger hormones.
Best of all, it felt so uplifting to have that control that, when I passed the car driver and his wife with their shopping trolley, I could look them in the eye and say 'good morning'.

Finally, a wonderful kindness to our self is to sit somewhere quiet, just for a few minutes and let go. Drop our shoulders, unclench our teeth, let go of our wrists and ankles, breathe from our tummies and ... smile to our self.

That simple act washes us with 'feel-good' hormones that can enhance a sense of peace and calm and is a step towards a truly profound new kind of being.

Now we're ready take the third step in Reversing the Loop of our negative bio-emotional response to stress.

Ready?

STEP 3

FEAR NOT

Fear is a terrible state of mind.

It is caused by extreme anxiety due to a dangerous situation, real or imagined, or sudden and unexpected, like an Earthquake, or waking-up to the sound of breaking glass.

Or by an on-going situation like living in a war-zone or bullying in the work-place, home or school.

Sometimes it is irrational, like phobias; logically we understand that a mouse cannot endanger our life yet someone who is fearful of mice will have a massive stress response if one scuttles across the floor.

Some fears are 'conditioned', an unpleasant shock can imprint fear into our very cells and we 'learn' to be afraid.

Here's a true story...

Spiders don't particularly frighten me but one day I was un-pegging socks from a washing line and a large fat brown spider scuttled from behind a sock, up my bare arm onto my neck then jumped off as I screamed and unpeeled myself from the chimney-pot.

From then on, whenever un-pegging clothes I did so with bated breath, expecting to see a huge spider and my body would prepare itself for flight.

I had become 'conditioned' to expect a spider lying in-wait on the washing line.

The thing is this – the fear response can often do us more harm than the actual danger – take my sock-spider, it was hardly going to kill me but I nearly had a heart attack.

The thing is this – humans have evolved to be afraid of things that 'scuttle' because in our long history of living in trees and caves our ancestors experienced pain and death from poisonous 'scuttlers' so fear of them became genetically programmed so we could react with reflexes faster than the speed of thought. I.e. logic doesn't come into it.

So, our autonomic reflex actions to fear are very primitive and undoubtedly save our lives when faced with sudden danger.

Sometimes, super-human feats of strength and bravery happen as a result but there's usually a pay-back …

Delayed-shock is a danger in itself and happens after a catastrophic event because at the time, our bodies reacted so fast we didn't emotionally process what was happening; that comes later when the immediate danger has passed and our emotions take-over, *now* we start to react with fear and shock.

Sometimes this develops into Post Traumatic Stress Disorder which locks the event into our 'cells' and becomes a permanent state of mind with physical symptoms and even a change in personality and for which a lot of therapy is needed.

Happily, learning to reverse the stress-fear-loops can also help with PTSD.

You may be wondering what all this has to do with Homo spiritus. You may be thinking are we to evolve into fear-less creatures, impervious to danger?

Far from it.

For as long as humans live on planet Earth, hurtling

through space, we need our fear-reflexes to ensure our survival; so why on earth would we want to reverse it.

Actually, we don't.

What we want to do is stop our ancestral bio-emotional loops getting the better of us when it's not in our best interest to react to an event with fear or anger.

What we want to be able to do is to control it so as to enhance our experience of life, and not just because we want to stop the situation getting worse but also because humans are a bit weird, we can actually invite 'fear' into our lives ...

Hey ho ...

Positive Fear v Negative Fear

Now, you'd never catch me tied to a length of elastic and jumping off a cliff – I'm much too frightened of heights; yet for others, bungee-jumping is exciting. It's a thrill, an adrenalin rush.

For them, this is a positive fear – yep, their pupils will widen, their heart rate increase and they may need to take some deep breaths but it's all worth it because it is something they want to do.

They choose this kind of fear.

Weird – I know.

If I was pushed to do it this is how my body would react:

Jelly legs
Dry mouth
Racing heart
Perspiration
Squidgy bowels

Tight stomach
Stop breathing
Sob

Not good.

It is astonishing how much positive fear we invite into our lives, in fact we wouldn't be very human if we didn't and once again it's all down to our ability to make choices because we are a species with free-will.

For example:

Taking exams
Performing on stage
Learning to drive
Competing at sport
Getting married
Job interviews
Meeting new people
Starting a new job
Travelling
Air-flights
Socialising

Okay, we don't *have* to do any of those things ... that's the upside of having free-will.

We could simply choose to become reclusive and turn our back on the world but most of aren't like that.

The thing is - Humans are genetically inclined to achieve, to push boundaries, to learn, to explore and discover.

Our happiness and sense of fulfilment comes from all these things but suffering stage-fright, exam/interview nerves, panic attacks, self-doubt, shyness and trepidation doesn't make it easy.

The weird thing is this, it doesn't matter if our fear comes from things we want to achieve ... a wedding; learn to drive; acting on stage ... or from the things we don't want in our lives ... a bully; a car crash; a stalker; bankruptcy ... our bio-emotional loops to fear are exactly the same and the effects are as follows:

Blood pressure rises.
Blood sugars increase
Bowels loosen
Breathing becomes erratic
Dizziness
Eyes widen
Fainting
Heart-beat quickens.
Muscles tighten
Perspiration increases
Saliva stops
Stomach acid increases
Stress-Hormones surge

Pretty impressive, huh?

But that's not all.
That can all lead to:

Acid reflux
Constipation
Cravings -food / drink /narcotics
Depression
Diarrhoea
Fog-brain

Headaches
Indecision
Indigestion
Insomnia
Irritability/anger
Lack of confidence/self esteem
Lowered immune system
Muscle aches
Night sweats
Palpitations
Panic attacks
Unhappiness

As Emotional Beings we sure don't have it easy.

Fortunately, once the fear goes away the body can start regaining its homoeostasis (balance).

If it was a negative fear the legs may wobble, tears may flow, and hopefully someone will give us a kindly cup of tea and a hug to get over the fright.

HOWEVER - If it's a positive experience i.e. something we invite into our life– such as stage-fright, then, once we're on stage the fear becomes a buzz.

After the performance the adrenalin shut-down still occurs but this time the tears are happy tears and the champagne and hugs are for celebration.

Remember, the fear response has evolved for sudden and immediate reaction to protect life and limb from harm or death, to keep us alert in times of danger – Earthquake! - MOVE! Enemy advancing! - FLEE or FIGHT! Prowling tiger – FREEZE!

Once the danger has passed we then collapse in a heap, and possibly weep, cheer, hug, eat, sleep and return to a nice period of enjoying a calm and productive life - until the next emergency.

The point is this: our fear response should not happen too often, not even the positive fear- too much of a good thing can be just as harmful.

Thrill-junkies beware!

So, how can we reduce our fear-response so life can be a bit more jolly?

Okay, In steps 1 & 2 we learned about the bio-emotional loops of anxiety and stress and how to reverse them by breathing correctly from our diaphragms (tummies); by dropping hunched shoulders; unclenching our teeth and, when really stressed, to smile a small Mona Lisa smile.

These small actions reduce the deadly effect of adrenalin etc.

In a sense we are 'kidding' our bodies that we're not *that* stressed or *that* anxious and that's good because it doesn't always help to get stressed.

Sometimes we need to stay cool and calm so we can handle the situation better and get better results and live life to the full.

The benefit of reversing these loops is that we feel so much more in control of our lives and it reduces the harmful effect of stress on our health, not to mention our relationships, work and socialising.

So what about the extreme kind of stress? What bio-emotional loop can we possibly reverse to help us cope with fear?

Take another look at the list of effects above. What, apart from forcing a smile and taking deep breaths, can also be reversed by our conscious will?

What else can we 'force' so our bio-emotional loop cuts out and we trick the body into believing that, actually we're fine sitting in the car taking our driving test - no sweat.

Can you spot it? ...

Fear creates a dry mouth

General anxiety can give us very unpleasant emotional and physical symptoms but when we realise our mouth is dry and it's difficult to get saliva to flow then we know our anxiety has notched up a gear – we are now feeling fear.

Take the stress of exams – most of us feel anxious when about to sit an exam; the adrenalin kicks in, our heart rate increases, our breathing quickens, our palms may start to sweat ... this is our body's normal reaction to anxiety.

Reversing the loops we've learned so far will help us keep our nerves both before and during the exam, however, if we realise our mouth has dried up then we know fear has kicked in.

This is when we need to sit and collect saliva in our mouth, as well as breathe deep from our tummy and also smile that tiny smile.

Now, there's a very good evolutionary reason why our salivary glands dry-up when we are fearful. It's because, as our body prepares to launch itself into fight or flight or freeze, the digestive system has to shut down, fast!

Digestion takes an awful lot of energy and effort; it's a massive bio-chemical process, best done when we're relaxed and at leisure.

When we need to remove ourselves quickly from danger our muscles take immediate priority, so blood, glucose, enzymes and hormones flood away from our gut and into our muscles.

Saliva production is the first step in our digestive process so when faced with fear the message is: 'Don't eat! You haven't time – get moving – keep alert – be prepared!'

The message is – 'prioritise! What's important here –

eating or escaping from this highly stressful/dangerous situation?'

Another fear induced response is the need to rush to the loo! Emptying your bowels and bladder completes the process of digestion. Once the gut is empty the body can flee and fight so much better. Anyone with irritable bowel syndrome or stress-incontinence knows all about this.

All animals do this – I had a little dog who hated being clipped; as soon as we turned into the street with the dog grooming salon he'd poop non-stop – like little bullets, and try to escape and then pee all over the salon floor.

Unfortunately I couldn't teach my little pal how to reverse his fear response, all I could do was cuddle him and give him treats.

Humans can; we can 'analyse' a situation; we can assess if it's life-threatening.

We can un-condition ourselves.

We can use our will-power – our ability to over-ride our bio-emotional loops so as to make life much more pleasant and rewarding.

Okay, so emptying our bowels and bladders in a crisis is not within our control, right?

Neither is our heart-beat or sweat glands.

But producing saliva is!

Picture the scene...

A basketball match at Uni, an important match between rival teams; the cheer-leaders are doing their stuff, there's a good crowd and the referee has his hands full as the two teams race around the court.

Two players are jumping high for the ball, shoulder to shoulder; there is a sudden loud crack as an elbow connects with bone and one of them lands on his feet with his hands over his nose; hands that quickly turn red with a gush of blood.

Feeling dizzy and in pain the injured player sinks to his knees. Players and cheer-leaders look alarmed and draw near.

The ref blows his whistle; the game stops.

The coach runs over with a hold-all.

'Luther? Let me see your face.'

A few minutes later Luther has been patched-up but the coach isn't happy. He looks at the tall, good-looking brown-skinned young man.

'Luther, your nose is broken. Now, I can take you to hospital or I can fix it here, but it will hurt like hell.'

People wince.

Luther nods but holds up a finger.

The coach waits.

Luther knows what to do; he stares into the middle distance, his mind focusing as he starts taking slow deep breaths. Slowly his shoulders drop, his face becomes calm but most importantly he moistens his lips with his tongue then starts to work the saliva into his mouth.

Everyone watches, puzzled and then Luther looks at the coach, smiles, nods then stares again into the middle distance, all the time working saliva into his mouth.

The coach leans in and in one swift movement with the edges of both hands he fixes Luther's nose.

The basket-ball player never flinches or makes a sound.

Everyone is impressed.

'I'm afraid you'll have two black eyes in a couple of days,' says the coach.

Luther stands up and grins. 'No worries; play on?'

'Not you, my son. Take this ice pack and watch, if you like, but this game's over for you.'

Luther is clearly disappointed.

It is possible to focus on your mouth, to work the tongue and the inside of your cheeks, to 'tell' the salivary glands to start working and to let the saliva build-up in your mouth – the key is not to swallow your saliva, the key is to keep your mouth moist, as did Luther.

Have a go, now, while reading this.

Forcing your salivary glands to work and to hold the saliva in your mouth reverses the bio-chemical loop of fear due to pain and something amazing happens – we feel less pain!

By holding saliva in our mouth the message is 'I'm not afraid, I'm in control, there's no need to fight, flee of freeze'.

At the same time, we breathe deeply through our nose, pushing-out our stomach to expand the rib-cage to get plenty of oxygen and drop our shoulders: unclench our teeth and smile that invisible smile that secretly empowers us.

Ever wondered why chewing peppermint gum helps concentration? Now you know.

Chewing stimulates saliva; saliva in your mouth is a signal that there's no need to get stressed-out so we stay focused and alert; The peppermint helps in the digestion – speeds up the break-down of fats and reduced stomach acid.

The message YOU are now sending to your body is one of control. It says - 'I can handle this' and your body listens and your emotional response listens.

You have reversed another ancestral bio-emotional loop.

Reversing these bio-emotional Loops of cause and effect doesn't mean that the 'bully' walking towards you - who makes your life hell at work, school or at home is going to turn into a nice and caring person, or disappear in a puff of smoke.

It doesn't mean that your teenage daughter who was not on the last bus home is safe and well.

It doesn't mean that mice will become your favourite furry friends; or air-travel your favourite mode of transport.

What it does mean is that the situation will not get the better of you and you will handle the situation without having a melt-down.

It means the mind/body loop is 'tricked' so that not every scary situation is treated as a matter of life and death. If it had been you wouldn't have had time to breathe, relax and pull saliva in your mouth. You'd have acted faster than the speed of thought.

If that happens and you start to experience delayed-shock, that is when you start to reverse the loop; that is when you breathe, relax all your muscles, get saliva in your mouth and try a little smile and make yourself a cup of tea (or a shot of brandy).

You will then find that the body's homeostasis returns quicker than you or anyone else would expect.

Reversing the loop can also help with flash-back ...

Flash-back is part and parcel of PTSD

Here's another true story...

Driving the car one cold winter's day I got into a bad skid on invisible black ice, skated across the intersection into on-coming traffic with my life flashing before my eyes – somehow I avoided a crash and ended-up buried in a huge drift of soft snow.

No-one was hurt; the car undamaged and some hunky truck drivers pulled over and dug me out. I carried on driving but had delayed shock two days later. I got over that but then started with flash-back, having nightmares of skidding.

Then I discovered how to reverse the Loops and by practicing these techniques every time I woke up after a nightmare, the flash-back and nightmares stopped, surprisingly quickly.

Pulling saliva into our mouth is not instant or easy but the more we practice the easier it becomes because pain isn't the only fear stimulant.

Picture the scene...

It's a busy train station in a large city, crowds are milling, trains are screeching and a young woman is darting along the platform, her rucksack bumping on her back and her face hot and red.

Kelly is a fledgling Homo spiritus, though she doesn't know it.

Several weeks ago she attended a workshop on Stress-busting with her friend Ellie and was taught about Reversing Loops. She isn't finding it easy and often forgets to practise but the knowledge is there.

Like everyone, Kelly has one or two bricks she throws in her way, to trip over now and again. One of these is a habit of being late. She leaves things to the last second.

Kelly misses her train.

She stands in horror and watches it pull away from the platform.

Her mind whirls in a panic; she has a connection to catch to get the last train back to Uni. She's frightened she'll miss it ... oh God! She'll have nowhere to sleep! Oh God!

Kelly's bowels start to do strange things and she want to burst into tears as fear overwhelms her. She's on her own and doesn't know which way to turn ... what can she do?

Who can she call?

Grabbing her mobile she calls her Dad but she hadn't had time to recharge it that morning, the battery is low and in horror she sees the screen turn black.

As Kelly's legs turn to jelly and she feels her mouth dry-up she remembers the workshop. She stands still, breathes deep and works hard to pull saliva into her mouth and starts to feel a little calmer and moves toward Train Enquiries.

And guess what?

Kelly is able to explain calmly and clearly what has happened; she's directed to a different platform, she will need to change trains twice, which is a pain, but she will catch her connection if she's quick.

Kelly does get the last train to Uni and sleeps comfortably in her own bed.

Just as important Kelly managed to take control so didn't have a major melt-down and even more importantly vowed to improve her time-management skills.

As we start to take control of our fear and stress levels we really do become more organised and efficient to the envy of all our friends and deeply impress our parents, family and partners.

Wow!

So what about Conditioned fear? Or fear that seems to have no logical explanation?

We can switch off our bio-emotional fear response

When we become conditioned to feel fear it is life-changing in the worst possible way.

There are a number of psychological techniques to desensitise us from phobias, obsessions etc and it's worth exploring these via your doctor.

However, it may be worth trying to de-condition yourself by reversing the loops as described in this book.

It worked for me.

You remember my sock-spider and how that *one* event conditioned me to be wary of un-pegging my washing?

Well, my mind (intellect) didn't really believe another large fat spider would 'attack' me but my emotions believed otherwise so every time I un-pegged the washing my body went into red-alert and I was always searching for spiders.

Finally I decided enough was enough.

This is what I did:

First, I just looked at my washing hanging on the line and, becoming aware of my tension, practised reversing the loops – the deep breathing, smiling, saliva, etc.

It was hard work but as I un-pegged the washing and no spiders appeared I felt all the tension ease and my smile became a grin.

After several weeks that fear had been banished from my bio-emotional conditioning, I no longer gave spiders a thought as I un-peg my washing.

For me on that occasion of de-conditioning it was the Mona-Lisa smile that really changed my fear-response.

I actually began to feel it that what happened was quite funny and began to adopt an objective rather than subjective view-point of the whole event.

WOW!

So what about pain and the fear that causes?

Without doubt pain, or the threat of it, is always a mega stimulant for fear, as we saw with Luther.

Having injections; having stitches taken out; the threat

of punishment ... wham! The fear response kicks in.

We begin to sweat, to breathe fast, our heart rate increases, our muscles tense, our mouth goes dry ...

Whoa!

Reverse the Loop!

Obviously if it's an aggressive or dangerous situation that's likely to cause you pain or injury the best solution may be to run and scream for help, if you can.

If it's a routine kind of situation like being at the doctors or in hospital then both before and during the pain-event keep pulling saliva in your mouth; breathe from your diaphragm, unclench your teeth, drop your shoulders and don't forget that Mona-Lisa smile.

There was a Spanish surgeon who removed varicose veins without anaesthetic by asking the patient to continuously produce saliva during the procedure. The patients had to practice this at home and have a friend with them who could help them focus on this but it worked time and time again.

WE can experiment with this: Talk it over with a friend then pinch one another on the flesh near our wrist, first without a build-up of saliva and then with.

We find that when we have collected saliva in our mouths before and during the pinch we will be able to take a much stronger pinch before feeling pain.

The anticipation of pain causes the fear response; remove the fear response and the pain signals to the brain are reduced.

As always, practice is the key to reversing these loops. It's not a one-off event.

With any medical procedure I always try to reverse all

the loops of breathing, smiling and saliva and generally impress the medics who have no idea what I'm doing and it also lowers my blood pressure.

As for bungee-jumping?

Not yet, but you never know ...

Lastly ...

Repetition is the key

The more we practice and repeat these techniques the more natural it becomes; in other words it becomes second nature not to throw an emotional time-bomb into a situation.

It becomes second nature to keep a cool head – no matter what our personality.

In fact, most personalities begin to improve and strengthen which is what Homo spiritus is all about.

It has a knock-on effect, too. As a parent you will be creating a new role-model for your children; also to your friends; colleagues and neighbours.

Quick note: This kind of change can never be forced or coerced on someone; it is up to each individual to choose for themselves.

This is a very quiet revolution, an inner revolution, a re-evolution of our core-behaviour.

To force or command this change on others would be the old Homo sapiens way which then leads to resentment, fear and anger and so the old loops of revengeful behaviour would continue, as History shows us in tragic abundance.

Homo spiritus understands this.

Homo-spiritus is not about controlling others.

Homo spiritus understand by not reacting to fear in the Homo sapiens way of knee-jerk panic, anger, greed or fear enables humans to behave in a calm, strong and gentle manner and then extraordinary things can follow.

Doors open.

We have a new-found courage.

We're better at decision-making.

We stop wasting energy and time on destructive habits and emotions.

People recognise our inner potential and our life become more fulfilled.

Now, imagine if everybody started behaving this way.

WOW!

It always starts with us, you and me.

When society changes it always starts with ONE person, be it in a family, a classroom; a street; a village ... a nation.

In Part 2 we will discover more about Fear and how this state of mind rules our existence and how it has evolved into something with extraordinary force. A force so terrible that has wreaked unbelievable damage on our psyche and on our societies.

It is the reason why so many of us are ill, threatened, at war, miserable, isolated, depressed and fearful.

It explains why so much of human behaviour defies common sense and is so frequently self-destructive.

It is time to change.

Maybe you already know what this loop is? Okay, then the jelly is setting nicely – but it's not ready yet, so don't peep.

We still need to reverse loops that challenge our physical well-being – loops of every-day existence and then it will be time to Reverse the most Powerful Loop of them all.

So, let the journey continue ...

HOMO SPIRITUS

PART 2

The Next 3 Steps

Mapping it out

So, we are now stepping lightly towards a leap in our 'spiritual' evolution, and not before time ... some might say.

Hopefully we've made a good start but there's still a long way to go. It is now the climb gets steeper as we move further away from our sea of troubles.

In fact, at times, it will seem we have a mountain to climb as we radically change the direction of our thoughts and our knee-jerk behaviour.

Homo sapiens have been trapped for thousands of years in ancestral cycles of self-destructive behaviour based on fear, greed and revenge.

The point is we don't need to live this way

We know this.

One way to fix this is to reverse the loops of our bio-emotional feedback systems - by breathing correctly; dropping our shoulders and unclenching our fists and teeth when feeling stressed or angry.

To smile when feeling anxious and when anxiety turns to fear we need pull saliva into our mouths and reverse total panic.

 (Unless we *need* to panic, then, trust me, you won't even think about reversing these loops - until after the event.)

Maybe we will start to cope with stress a bit better, feel calmer and be better organised; maybe our decisions will be wiser as we become more thoughtful and considerate.

Maybe our relationships will improve and with it our self-confidence and self-worth.

Great!

If this starts to happen we will surely be on the path to becoming a new kind of human, though it will take time, maybe a life-time. The trick is, never to give up, not even when we forget to reverse the loops which can happen daily.

We will still need to let off steam, on occasion; and feel anger, resentment, impatient etc. The trick is to temper them and then un-wind them and then look at how the situation could have been handled differently.

Does this mean all our problems are solved?

Hardly.

Reversing our bio-emotional loops protects us from the harm stress can do to us but it doesn't remove the stress from our lives.

Knowing and understanding why we do the things we do is the next part of our journey.

Picture the scene ...

It is 7.30 and dawn has broken over a city sky-line; the traffic is building-up and an alarm rings at the bedside of a fledgling Homo spiritus.

Ellie has moved out of her Mum's house and lies crumpled under her duvet. As her eye-lids peel open she gropes for her mobile and cancels the alarm. Ellie feels exhausted, she didn't sleep well ... she never sleeps well, these days.

Sometimes she thinks it would be nice to go back home again, to her Mum.

With a sigh she stretches and as sleep fades from her mind a feeling of dread takes its place.

It is a work day.

She can't lie in bed; she has to get up and pushes one leg out from under the duvet.

Ellie hates her work, hates the dull routine, the management; the complaining customers, the stress of meeting targets, the back-stabbing; the lack of natural daylight ...

Dragging herself out of bed she opens the curtains, the sun is rising but Ellie feels no joy at the prospect of a new day. The muscles tighten across her shoulder blades; a 'stone' settles in her stomach and gritting her teeth she staggers into the bathroom.

Ellie stares at herself in the mirror and feels upset at what she sees. 'This isn't me', she thinks.

Her mouth is as dry and her tongue seems swollen, she takes a drink of water and starts removing yesterday's make-up. Her mouth soon feels dry again.

Ellie has attended some of Suzi's workshop and knows about the bio-emotional loops and realises what her body is telling her – her dread of going to work has kicked-in a fear response.

For all the world she'd stay in her bed-sit and pull a sicky, but that really isn't an option.

Sighing deeply she stares at her reflection unclenches her teeth and starts pulling saliva into her mouth and does her best to smile and breathe deeply.

It isn't easy, the smile is wobbly but she perseveres and returns to her bedroom to get dressed.

Something extraordinary starts to happen, as she prepares breakfast - chocolate spread on toast and a mug of coffee, Ellie no longer focuses on the dread of going to work; random thoughts drift through her mind instead

It is as she locks the front door behind her that the feeling of dread begins to return and Ellie realises, with hindsight, how her fear had subsided for a short while.

So she continues reversing her body's reaction to stress on the way to work – forcing herself to smile a tiny smile while sitting on the bus … walking into the building … saying 'good morning' to people … … taking off her jacket … getting a coffee …sitting at her desk, switching on the computer.

'Okay, I got here,' she thinks. 'But this isn't what I want to do for the rest of my life.'

Then her emails start downloading, dozens of them.

She sighs and rubs her neck as knots of tension take up position, if only she didn't feel so tired all the time. Pulling open a drawer Ellie takes a toffee from an opened packet and pops it in her mouth then starts on the e mails.

So, Ellie was able to get to work without a complete meltdown, without raising her blood pressure or giving herself a migraine but it wasn't easy and she still hates where she works and there are still plenty of niggles in her life ... sleep, diet, not feeling totally in control ...

If Ellie continues reversing the bio-emotional-loops of stress, her life will slowly change but there is so much more she can do for herself.

Knowing she wants to change her life is one thing, doing it is another.

That is what Part 2 is all about.

We now start to take a broader look at our lives and the loops of behaviour and conditioning that can trap us into cycles of misery.

Such as:

- Why don't I sleep well?
- Why am I putting on weight?
- Why am I in debt?
- Why am I in a job I hate?
- Why are my relationships struggling?
- Why do I feel restless?
- Why do I ache so much?
- Why am I on anti-depressants?

- Who is the real me?

Homo Spiritus needs answers to these questions before we can walk on this planet with a lighter step, shoulders back and chins lifted.

As always it's best to take one step at a time, the jelly is setting nicely but slowly. Leap to the middle or the end and it won't hold its shape and if you've ever tried eating an unset jelly ... well, it's pretty hard to swallow.

It takes time for an idea to grow; evolution takes even longer but in our case it doesn't have to take eons because we can engineer our own future. We do it all the time and have adapted the world to suit our needs, remember? – Cut tunnels into mountains; drained marshes; cut down trees; built cities, harbours ...

Intellectually we are evolving, too; more and more societies no longer believe they are puppets of the Gods or slaves to superstition. Self-determination is slowly emerging for women and men and it is a wonderful thing.

Or it should be.

The thing is this:

If we continue to live our lives as 'victims' of our own self-destruction, we are not going to be well in mind body or spirit. And if we are not well in mind, body and spirit then no amount of self-determination or mastery over nature can make us happy.

It's time to take the next step towards changing our future. It is time to reverse the loops of behaviour that cause us sickness in mind, body and spirit.

Ready to get better?

STEP 4

BE WELL

Feeling well is a state of Mind, body and Spirit. It means being content with our life, to be free from pain and discomfort and it means being comfortable in our own skin, happy with who we are and where we are going.

Now, that seems like a tall order, so how do we get there? One step at a time, that's how.

Okay, so we know that breathing is essential to life and that for humans breathing can become dysfunctional because it is intimately linked to our emotions, to our muscles, to our hormones - to how we are *feeling*.

We also know that not getting enough oxygen makes us tired, achy, fog-brained and the rest! – a vicious loop that we can, if we chose, reverse because we have free will, and by so doing we are able to restore some harmony to the mind and body

But what about the 'spirit' – that hard to define 'something', that gives each and every one of us a unique experience of life? To feel 'low in spirit' is something most of us have experienced and sometimes the feeling can seem to last forever.

Are there any more Loops that need reversing, loops that can bind us to a life of ill-health, struggle and misery and which can defeat our spirit?

Oh, you bet there are but we can, if we chose, reverse these evolutionary loops too because we are so very, very clever.

All it takes is a new kind of awareness and then the deliberate decision to go for it!

So, let's start Step 4 with something we spend one third of our life doing.

Sleep Well

Sleep ... nature's soft nurse ... (Shakespeare)

Ellie wasn't sleeping well for lots of reasons, but basically Shakespeare got it right – sleep is nature's nurse and it seems that all life needs a period of rest in every twenty-four hours. Even plants rest at night – petals close and leaves rest as they no longer have to track the sun.

How very lucky life is, on planet Earth, to spend half it existence in darkness – away from sunlight, the very thing that gives life to our planet home.

Hmm.

Makes you think.

Makes me think.

Something is going on here. Something very important and of course luck has nothing to do with it. The bliss of darkness, of rest and stillness – is this too an Essential for a life? Does life need long periods of darkness in order to exist?

Has life evolved because there are both periods of light and dark?

It would seem so.

Sleep is certainly something of a mystery and why we humans sleep away one third of our lives in a state of deep unconsciousness and the land of dreams is an even greater mystery.

However, as with all mysteries, clever Homo sapiens is unravelling its secrets and if you're interested in the different cycles of brain-waves during sleep, dreaming, and the changes in our muscles, hormones, breathing etc then the Internet has up-to-date information.

It will tell you, for instance that different species of animals need different amounts of sleep, from snoozing for a few hours a day like a donkey or simply 'resting' like a fish.

It will tell you that sleep and dreaming is essential for brain function, development, memory and learning. And it would seem that the bigger the brain relative to the size of the species the more sleep it needs.

It will also tell you that long term sleep deprivation causes hallucinations, impaired speech and comprehension, lack of muscle coordination, personality breakdown and will eventually lead to insanity and death.

Hmm, something is definitely going on here.

Clearly sleeping is as fundamental to life as is breathing and, just as humans can become dysfunctional breathers can we also become dysfunctional sleepers?

The answer is – yes!

Chronic insomnia is a curse for many people.

Just to have a broken nights' sleep is a nuisance and can lead to health niggles such as exhaustion, headaches, irritability, poor concentration; lowered immune system and depression which can then lead to more health problems.

And guess what? That little lot can seriously disturb our sleep pattern in the first place. Do we have another loop here?

We certainly do! Good health and peace of mind is essential for good sleep and yet a good night's sleep is essential for good health and peace of mind.

Cause and effect again. Loop-de-loop-de-loop.

The issue is made a lot worse because, having an awareness of Self we know we have insomnia and can dread going to bed knowing what kind of a night is in store for us.

We know we are awake when we should be asleep and this awareness starts a cycle of worrying because we're not sleeping and not sleeping because we're worrying and the loop goes round and round and round.

So, let's have a closer look at the possible reasons for dysfunctional sleeping and other facts ... most of which you probably know:

- Not everyone needs the same amount of sleep.

- Our sleep requirement changes as we get older.

- We can get into a *habit* of waking up during the night.

- If we are cold, too hot, hungry or thirsty we may not sleep well.

- If there is too much noise or too much light we may not sleep well.

- If we eat or drink too much on an evening we may not sleep well.

- Several food and drink substances are stimulants may keep us awake.

- Too much mental stimulation before bedtime may keep us awake.

- If we are in pain or discomfort we will not sleep well.

- If we have unfinished business we will not sleep well.

- If we are worried about something or have an over-

active mind we may not sleep well.
- If we dread the morning we will not sleep well.

Okay, the golden rule is this, if we wake-up every morning feeling refreshed, wide-awake and longing for the day to start then we're not a dysfunctional sleeper.

In fact, very likely we're not a dysfunctional anything.

You will see from the above list that several reasons for poor sleep are to do with our physical comfort and these are relatively simple to correct.

Some of the solutions are pretty obvious and my work has taught me the rest. Some of the solutions seem a bit weird but hang-in there; these have been tried and tested.

Of course, if you have a health problem that is keeping you awake you must consult with your doctor.

- Change your mattress and pillows as they become unfit for purpose.
- If a habit has developed for waking-up during the night, e.g. exam revision; having a baby; try a course of relaxing herbal sleep remedies, take them for several weeks then stop and see if the habit has broken. Do check if you are taking other medication.
- A drink of warm milk or a camomile tea at bedtime. If you have no trouble falling asleep but wake a few hours later save your chamomile tea for then, have it by the side of your bed and drink it cold.
- A bowl of cereal at bedtime.
- Wear bed-socks.
- Use ear-plugs and/or eye mask to cut out noise or light. A cotton scarf tied over your eyes and

ears may be more comfortable and probably entertain your sleeping partner.
- Use dim lighting as you get ready for bed – undressing; removing make-up, brushing your teeth - the reduced light signals the brain (pineal gland) it is time to prepare for sleep. *(By the way, the pineal gland is totally amazing, about the size of a pea, it produces hormones and is wrapped in mystery. Something else to search on-line)*
- To still the mind practice a relaxation technique once you've settled down to sleep. Focusing on breathing from your diaphragm and unclenching you teeth and muscles maybe all you need to do.
- Some people find reading or listening to the radio or to some music helps send them to sleep.
- If you wake-up in the middle of the night avoid getting out of bed if possible and continue with the relaxation technique or switch on the music again.
- Take some pain-killers at bedtime (consult doctor) if you have aches and pains.
- If you are likely to wake-up with pain in the middle of the night, have a soluble pain-killer by your bed-side ready to drink. Avoid putting on a bright light. Use a small torch if necessary.
- Avoid looking at your mobile phone or other illuminated screen.
- For cramp try a small drink of tonic water before bed and have a couple of corks wrapped in a rubber-band in the bed, close to your feet.
- Avoid caffeine after 14.00.
- Avoid eating your evening meal after 19.00
- Avoid rich and spicy food, including garlic,

chocolate and cheese on an evening.
- Make a list of all your unfinished 'jobs'. Number them in order of priority then put the list on the fridge door or kitchen work-top – outside your bedroom door, not IN your bedroom.
- Make your bedroom a haven for sleep – no desk, TV. computer, they belong in the study/living-room.
- Insomnia due to worries, anxiety, grief, shock and fear can also be helped by Reversing Loops 1, 2 & 3.
- My favourite solution for getting back to sleep is to get really comfortable, relax all my muscles and then 'smile' myself back to sleep. It takes concentration and patience but can be very powerful.

In addition to that little lot, longevity and good health seems to be closely linked to the quality of our sleep and our brain-function and those clients who are in their eighties and nineties and who are remarkably well, tend to be living proof of the following 'old wives' tales' concerning sleep ...
1. Early to bed, early to rise makes a soul healthy and wise.
2. One hours sleep before midnight is worth two hours of sleep after midnight.
3. A small shot of whisky, rum or brandy in little hot water/milk at bedtime can be helpful. If appropriate.
4. Procrastination is a thief of time and will steal your sleep.
5. Don't go to sleep on a cross word. Make your peace.
6. When in bed keep your feet warm and your face cool.

Go for it!

Okay, it may seem like that's a lot of changes to make but is it any wonder sleep can go so wrong?

The secret to success is so to let your sleep dysfunction correct itself gradually and naturally, one step at a time and not to try too hard.

Sometimes we just have to keep the **Intention** of falling asleep in our mind and then let-go of worrying about it, a sense of failure does us no good at all.

Intention, by the way, is a very powerful tool and well used by Homo spiritus. Most of us have something in our personality that trips us up, like Kelly and being late all the time, which then may disturb our sleep patterns and just having the Intention to 'change' can help.

Picture the scene ...

Kelly is running up the steps to the dentist's surgery; she's two minutes late for her appointment. She flies through the door and apologises to the receptionist.
'Hi Carol, sorry I'm a bit late.'
The receptionist smiles at her. 'It's the earliest you've ever been, Kelly.'
Kelly grins. 'I know and I'm much more organised these days and I'm sleeping better too'.
'That's wonderful,' says the older woman, giving Kelly a thoughtful look.
'Yeah, I've started doing Affirmations.'
'Affirmations?'
'Yeah, it really works. I tell myself, 'I am an early person and that I love being on-time.'
Carol laughs. 'That sounds a bit too easy.'
'Well, it's all part of my Intention Technique ... to change my

old habits. You've got to let go, and breathe correctly and smile.'
'Sm ... smile?' Carol was more than a little puzzled.
'Absolutely, I've been going to these workshops, they're ace. Oh and they really help with phobias and stuff. Here, I'll give you some of Suzi's cards to put out for your patients, you never know ... '
Kelly removes some cards from her satchel and puts them on the counter then takes a seat and closing her eyes, drops her shoulders, smiles a tiny smile and tells herself she likes going to the dentist and taking good care of her teeth.
Carol watches her for a moment then picks up one of Suzi's cards.

The reason Intention is so powerful is because it involves our sub-conscious mind which, while not something to be messed with, can also be our greatest ally.

In fact, the sub-conscious part of our mind is so 'tricky' we need to talk about it – in small bites.

Sleep & the Subconscious

Firstly, the sub-conscious mind takes on-board our wishes and desires, as well as our fears and insecurities.

It takes on board our belief-systems about ourselves, often 'suggested' by other people – parents, siblings, teachers, friends, partners, who may 'tell' us we are lazy, stupid, boring ...

Hey, come on! Haven't we all done exactly the same thing?

It's what Homo sapiens do – we make 'throw away' remarks all the time. Usually quite unaware of any 'harm' we are doing.

Okay, so we develop 'weaknesses' and we also get very good at pretending 'I'm fine'.

But not so the sub-conscious mind. Oh no, our little fiend, oops 'friend', knows when we're hiding something.

So, when we go to bed with any 'unfinished business' it is our sub-conscious that shouts at us to 'get up! You can't sleep; you've got stuff to do? Get up!'

Of course it's easy *not* to listen to our sub-conscious mind, that's why it's 'sub'-conscious, it's below our everyday level of consciousness, however, it is relentless and unforgiving.

And yet - because our sub-conscious mind can be so easily 'programmed' we can turn it to our advantage.

You start telling yourself you're an early person and sooner or later your subconscious takes this on board, begins to believe you are an early person and transformation happens!

You start behaving like an early person.

Intention works in many ways; if you feel you lack confidence and you start telling yourself you have lots of confidence and enjoy knowing your own mind then - hey – you become that person.

(There are a number of therapies that can help strengthen Intention and a list is at the back of the book.)

So, how does this help with our sleep problems?

Well, by simply having the Intention to sleep well and to believe you are an excellent sleeper and to say this to yourself while lying in the dark, letting go of all your muscles, breathing from your tummy and smiling a tiny smile can work wonders.

My favourite affirmation when my overactive mind keeps me awake and I've decided it is time to get back to sleep is to say to myself: 'Night is for sleeping; day is for thinking.'

It works a treat.

If it's Anxiety that keeps us awake then pulling saliva into our mouths as we repeat an affirmation, such as: 'I am safe and relaxed and things will get better', as well as relaxing and breathing correctly can really help.

By focusing on these tasks it takes our minds off all our worried thoughts, as well as switching off the fear response so we may fall asleep quicker than we realise.

In fact, reversing the loop of insomnia creates a completely new loop, one that is excellent for our well-being:

I worry so much that I don't sleep well.

By relaxing, smiling and breathing, I sleep better.

By sleeping better I don't worry so much.

We change.

Switching off the bio-emotional responses to stress and worry as we lie in bed transforms our sleep which in turn transforms our lives.

It takes time for this process to become a new 'habit' but when we sleep well the brain is rested; our hormones are regulated; stress melts away from our muscles and stomach.

To wake up in the morning feeling refreshed and rested really does make life easier to cope with.

The transformation to Homo spiritus changes us from the inside until one day we wake up and realise our old 'habits' of being late, of worrying, of not sleeping well, of nagging, denial, blame etc are starting to fade away.

When this happens we know the transformation is happening at a sub-conscious level and this is a huge step forward.

What we have been consciously trying to do is now starting to become second-nature!

However, supposing your nights are broken due to

circumstances beyond your control – a crying baby, a sick relative, arthritic pain, noisy neighbourhood …

What if your night-time is disturbed by conditions beyond your control so you inevitably wake-up feeling tired?

Well, here's a solution that may help …

Rest the Brain

Sleep is all about resting the brain; allowing the body to renew itself; letting the immune system do its job; laying down memories; allowing the sub-conscious to have its 'say' … and more.

But we don't always need to fall unconscious to rest the brain.

Meditation is an excellent way to rest the brain and it need only take a few minutes – in fact, a little and often is the best.

Now, before you're tempted to skip this page, thinking Meditation is not your cup-of-tea, hit the pause button because it's a lot easier than you think and for Homo spiritus meditation or resting the brain will be as normal as … breathing.

Meditation is a wonderful tool for helping the brain to 'sleep' and it doesn't have to be a complex or structured procedure.

It doesn't have to 'belong' to any religion though it may be useful to try a Meditation Group and find one that suits you and to help develop the practice.

Simply having the Intention to rest the brain and then to sit somewhere peaceful and to relax every muscle and to breathe from your belly and when a thought wonders into your mind, acknowledge that thought then smile and return your focus to the rise and fall of your tummy.

It's a simple as that.

Meditation is about focusing your mind on one 'thing', in this case your breathing, or it can be one word, one sound or a chant; the sound of waves; or a candle flame or merely gazing into the middle distance as in a day-dream.

It is also another example of how 'trying too hard' can be counter-productive.

Be warned, working too hard at 'stilling your mind' can become stressful!

The trick is to have no expectations and simply ...

Let-go.

Have the Intention and then let-go.

Picture the scene ...

The airliner is flying at thirty thousand feet, above the clouds. Far in the West the sun is setting where the sky is a canvass of a blue sky streaked with orange flames.

The plane is full. Every seat is taken, people shift, babies cry.

There are six hours of flight-time left.

For all his experience of flying and his rise as a homo spiritus Luther can never enjoy the experience of being in a plane. He's tired but his long legs and general discomfort means he can never sleep.

Sitting by the window Luther sighs and hunkers down to endure the long flight.

Two hours later he is at his worst and needs to walk about but the passengers next to him are fast asleep.

Lifting the window-blind he stares into the night sky. Meditation will help his state of mind.

So, slipping into de-stress mode Luther gazes deep into the distance. The sky is dark now, with a twinkling of distant stars; he empties his mind as a he concentrates on allowing his arms and wrists to lie slack on his knees, his ankles to release, his jaw to slacken and his breathing to become slow and deep.

He focuses on his breathing, allowing his thoughts to drift and then to return to his breathing. Scanning his body he checks from time to time that he is still fully relaxed, letting go of muscles that have tightened and with each out-breath he lets go that little bit more.

Luther's weight sinks deeper into the seat and yet his mind seems to expand outwards, towards the stars, into the darkness. As though in a day-dream – his brain waves slowing, in tune with his breathing. Slower, deeper.

The rattle of the hostess trolley catches Luther's attention. He blinks and stares at the steward who asks if he'd like a drink.

Luther has some water and realises he feels completely rested and relaxed, another hour had passed as though in a flash and it will soon be time for breakfast.

Now here's the important bit ...

People who meditate find that they need less sleep on a night.

Meditation alters the brain-wave frequencies and is nourishment for the brain. Try it, ten minutes is all it needs, even five minutes can work wonders and often feels longer.

Meditation quietens the endless 'chatter' of our thoughts and de-clutters our mind and when this happens, inspiration and intuition work much better for us 'thinking, knowing' humans and these are very powerful tools for Homo spiritus.

The good news is, by following the First Three steps in Part 1, we may find that our sleep patterns have already started to improve due to stress hormones no longer constantly flooding your body.

It's quite simple really:

A calm mind is essential for sleep. Sleep is essential for a calm mind.

The Loop has already started to Reverse and now, by checking our sleep isn't being interrupted by any of the above physical, emotional or personality hiccups and by using the power of Intention our sleep patterns can be transformed.

By the time we have finished reading this book and started joining up all the dots we may find that our sleep pattern has evolved into something shorter, deeper and infinitely refreshing and healing.

As a bonus, think of all the things we can achieve by having more wakeful hours by not needing so much sleep. Daytime will then become more enjoyable as we think clearer and make better decisions and thus our sleep gets better and better.

It's very powerful loop indeed.

So ... we're learning how to cope with stress; we're learning how to improve our sleep pattern – all essential tools for a healthier, happier life.

What's next? Well ... it's another biggy; it's also multi-layered so is not an easy loop to Reverse ... it's about Homo sapiens and our relationship with FOOD!

Eat Well

Food, glorious food!

Yep, we humans love food and we particularly like variety. ..

When you think pandas live off bamboo and koala bears live off eucalyptus leaves and even our closest cousins - chimpanzees and other primates – exist chiefly on leaves and fruit, it's a bit of a mystery why humans crave such a wide variety of tastes, textures and types of food.

Food manufacturers know this and capitalise on the fact

and are some of the biggest and wealthiest corporations on the planet and are forever tempting us with yet another variety of breakfast cereal, chocolate bar or fizzy drink.

The truth of the matter is Homo sapiens are addicted to food and our appetite for variety seems to know no bounds and it's not just variety but also the amount of food we consume.

The awkward fact is that if a stomach (essentially an empty bag) stretches because of the amount put inside, then it will take more food to fill it! And it is when the walls of the stomach 'detect' food that the brain gets the message we've had enough to eat.

A bigger stomach demands more food, more food is consumed so the stomach gets bigger ...

Another loop-de-loop-de-loop.

One way to reverse this loop is to make the stomach smaller and surgery is now used to achieve this as in gastric-bands.

Another way is to simply cut-down on the amount we eat but this is so very difficult. Why?

Not knowing when sufficient is enough or when enough is sufficient seems to be a problem for those of us who can afford to eat whenever we want and in the modern, so called 'developed' countries; hunger has little to do with the amount we eat.

So, why is it Homo sapiens love to eat?

Many people plan their lives around meal times, and food is their greatest pleasure. Nothing wrong with that but what happens if we become obese and then develop health problems; poor sleep patterns, low self-esteem and social difficulties as a result?

Why is our will-power so weak when it comes to food? Is there something stronger at work here ... stronger even than our own will-power?

Yep and it's called the Limbic System.

Hey ho ... here we go ...

The Limbic System is very rarely discussed, which is a pity because it basically explains why Humans are so emotional. Now, this too is something you can search online, it's ENORMOUS, complex and utterly fascinating.

Let's try and put it in a nutshell ...

The Limbic System is responsible for the Pleasure-Reward Response in humans and other mammals and it can create addictions! Ever had a pet who insists on being petted?

When we, and our pets, feel' pleasure' a rush of 'feel-good' hormones floods into our brain and we feel ... great!

Cats purr, dogs wag their tails and if a monkey gets a peanut for performing a trick, guess what? He'll perform as often as you wish ... the peanut, and hence pleasure, being the reward.

It's obvious, isn't it, that if animals, us included, didn't get pleasure from eating then maybe we'd not bother and then - well ...we'd become malnourished and starve.

We have evolved to like food.

Hunger pains are unpleasant – food takes away that discomfort – problem solved, but what happens when we eat when we're not hungry – and why would we?

For Homo sapiens the Limbic System has reached new heights. We have a sense of Self, remember? We are so *aware* of how we are feeling this increases our mental state whether it's a sense of anguish or euphoria.

Certainly, other animals are aware of hunger, pain, cold etc but we are *aware* of being aware. It's a whole new ball-game.

This is how and why something like a beautiful sunset or music or laughter can uplift our spirits and fill us with such a great sense of joy and wellbeing.

The opposite is also true, we only have to witness or experience all the suffering that Humans or animals endure to feel a crushing despair, grief and a sense of hopelessness.

You remember in Part 1 we talked about the Chinese concept of Yin & Yang – opposites as observed in the natural world? Well, the opposite of Reward is Punishment and guess what?

Cunning Homo sapiens are experts as using this as a form of control, we all do it, all the time, consciously or subconsciously.

If you're 'good' you get a reward if you're 'bad' you get punished. Children are allowed to stay up to watch T.V or go to bed with no pudding; gold stars at school or go stand outside the door; sweets or no sweets; praise or criticism; a kiss or a cold shoulder. And of course the ultimate threat - Heaven or Hell.

So, what has all this got to do with food?

First, let me tell you about a wonderful little old lady who, as I write this is nearly ninety-four years old and is still going strong. Her weight has never changed and she takes no medication and blood pressure is that of a thirty year old.

Now, Betty was sixteen years old when World War 2 started and she spent the next five years working in a bank, putting out fires, knitting socks for the army and living off very meagre food-rations. Her meat ration for a week was three small lamb chops, one egg, 2 ozs of cheese ... in fact, food rations continued until 1953.

'Did you go hungry?' I once asked.

'Not really. Everyone turned their lawns into vegetable gardens and while helpings were small we did eat three times a day.'

'It must have been miserable.'

'Well, I did miss oranges and bananas but honestly we never thought about it too much, because, you know, darling, we had much more important things on our minds.'

I stared at my mother and thought, 'yes – of course you did.'

Here's another true story. Once upon a time, and not so long ago, lived a lady who devoted her life to her horses – she was reputed to exist on black coffee and chocolate digestive-biscuits. She spent every day at the stables, in the open air, was never known to be ill and worked until she died in her ninetieth year.

There's a clue here.

My mother's desire to 'help win the war' and the horse-lady's desire to be with her horses were greater than their desire for food. Their Pleasure Reward System was 'fed' and met through a pair of knitted socks ... the birth of a foal ...

The Limbic System that demands the gratification of joy was 'fulfilled' not by putting food into their stomachs but by living a purposeful life.

Also, maybe, when living a life true to your Self the body takes it's nourishment from something other than food?

My mother adopted a frugal diet the rest of her life. Also, she has a terrific sense of fashion and being less than five feet tall understands that if she puts on weight she'd struggle to find the clothes she likes to wear. Clothes gave her more pleasure than stuffing her face.

Yes, she loves Madeira cake and dark chocolate (with sea-salt) and has a little whisky before bed and eats three meals a day but her portions are small. A bar of her favourite chocolate lasts a week.

She also loves life and people, history is her passion as is music from the 40's, not to mention her children, grandchildren, great grandchildren and the memories of her wonderful husband.

The thing is this ... when your life feels fulfilled, you can almost forget about food, until a loud and startling tummy rumble echoes around the room which then makes you aware you're hungry and it's time to eat.

The question then is ... what to eat?

There is enough information out there regarding weight loss, diets and nutrition to fill an Olympic stadium. It doesn't need to be repeated here except for one thing .... Ready?

It's what you have to *avoid* eating that is vitally important and there is one big fat bogey food that should be avoided like the plague ... like your worst nightmare ... like your wicked step-mother ...

SUGAR!

Sugar behaves like an acid and it probably has killed more people than the plague and is certainly a poison; it rots your from the inside out; it has no nutritious value, even bacteria can't live off sugar which is why it never decays.

Sugar works with your Limbic System like a drug, the more you have the more you want and you get desensitised to it.

Sugar is a quick limbic fix and that's not good news. It's a cheat, a swine and it's deadly!

Got the picture?

Let's go back a few hundred thousand years ...

Sugar wasn't around when Homo sapiens first walked up-right on the planet. Fruit and an occasional treasure find of honey was all the sweetness they experienced... and no doubt enjoyed! Mammalian milk too is rich in natural sugars for the newborn infant, yum!

Sweetness is a nice taste and natural sugars from fruit

and root vegetables and a little honey give vital energy. But processed sugar, refined and white and useless is 'empty calories' and our bodies don't know how to handle it.

You see, our physical evolution is too slow to adapt to this poison – and so sugar is turned to fat and destroys the pancreas in the process and yet food manufacturers add it to processed meat, pizza, fruit yoghurts ... Why?

Because it is addictive! They know their products will sell if sugar is added.

Don't fall for it. Check the labels, the first ingredient on the list is the largest quantity and if you see sugar or syrup, or dextrose or fructose anywhere near the top then put the item back on the shelf.

You know ... People have Power - if the stuff stops selling, the manufacturers will stop producing it.

Now, back to the Limbic System:

The thing is this ...sweetness and yummy food has become an affordable treat, a treat we now take for granted and have become addicted to.

Food is the Reward – the pleasure we give ourselves and it seems we need it more and more ... why?

Psychologists use a term Compensatory Behaviour (plenty about that on the internet) which refers to human behaviour in response to anxiety-causing problems.

Has life, for so many of us, become so un-rewarding we have turned to food as a compensation?

Has a high level of stress become the norm and so eating helps alleviate feeling stressed?

Of course it has.

As far as the Limbic System is concerned the only anecdote to stress is pleasure (more of this later), and while Reversing the Loops in Part 1 helps to negate the effect of stress on our health it doesn't fix the problem of having stress and what we do as a consequence.

Some of us smoke, shop, drink coffee, experiment with recreational drugs but the most common is to eat sugary, fat laden food. Tasty food is a cheap quick limbic treat but it is short-lived so we need more and more.

As a result we can become grazers, munching all day, feeding our Pleasure-Reward Limbic System with short but constant 'highs' to help get us through the day.

We can become fridge-raiders at two o'clock in the morning. Hmm.

We no longer eat because we are hungry ... we eat because we are bored, stressed, lonely, fed-up ... thus creating cycles of Self destructive behaviour. We binge so then we feel guilty so we feel depressed, so guess what ... we eat more.

Okay - now you're thinking ... if the Limbic System is stronger than our will-power how on earth can we reverse this loop?

Because, let's face it, diets fail. The Limbic System doesn't thank us for denying ourselves pleasure so eventually we 'give-in'. Also, following a diet regime often takes-up our lives so we seem to have little time for anything else.

(Now there's a clue for what's to come.)

Unless Homo sapiens (including food manufacturers) can find a way out of this situation, as a species we are very likely doomed to get more obese every generation, needing more and more medical intervention to keep us even half alive.

Not good news (except for the pharmaceutical companies.)

While we digest that unhappy scenario and before we

examine how Homo spiritus solves this problem, let's look at some useful tips & suggestions to help us eat better because being aware of some basics is part of the journey towards good health.

1. Only put food in the shopping trolley that is not processed or contains added sugar.
2. If junk food is not in our cupboards we can't reach for it.
3. When a chocolate-jag hits, try a chocolate drink instead, make it from raw cocoa powder and hot water then add milk and brown sugar, just enough to give it a pleasant taste and texture.
4. Reduce the size of our helpings by serving food on a smaller plate
5. Wait until we're hungry before eating.
6. Chose fruit/cheese or a delicious coffee instead of desert.
7. Eat regularly, don't skip meals – this will prevent binging.
8. Eat snacks that are only wholesome and fresh – nuts, fruit, and cubes of cheese.
9. Limit ready-made meals and 'take-aways' check for fat, sugar and MSG content.
10. Enjoy learning to prepare and cook food, on our own or with partner and children.
11. Get into the habit of saying 'no thanks' when someone offers you a sugar-laden, fat-rich treat.
12. Search for and keep an eye-out for healthy eating articles and recipes, make-up your own.
13. Try new taste sensations – like mashing a ripe avocado with ricotta cheese and dipping with carrot sticks (add a dash of

chilli or Worcester sauce).

Okaaay …. What you're thinking now is, doesn't that list involve Will Power? And didn't we just read that our Will Power is weak compared to our Limbic System?

Yes, but don't forget that other powerful tool that plays a huge role in the life of Homo sapiens, it's been mentioned before …

The Sub-conscious Mind

Our sub-conscious works in mysterious ways, even if you think the 'eat better' list above is just one mountain too steep to climb; a few suggestions will take seed.

It's all about having the **Intention** to Eat Well and to keep reading that list again and again, that way the sub-conscious will take on-board a few more suggestions.

Read the list out loud and the effect is even stronger.

It's not about *trying* to remember, as in parrot-fashion, it's about gradually assimilating new notions, new ideas … a new mind-set that our subconscious adopts over time and then becomes programmed into our very subtle genetic code.

The truth is, if something is repeated often enough it becomes 'embedded' in our behaviour or our belief-system. This is how the Power of Intention and also 'conditioning' happens and it doesn't always have to be repeated that many times. It depends on how strong our Intention is.

If we tell ourselves every day that we like being slim and like to eat well, then our sub-conscious will start to believe us and our sub-conscious always fulfils our beliefs.

Tell a child that he is a 'slow coach' and he will become that person; tell a child she is untidy and she will become that person. Tell someone they are stupid and they will believe it.

The sub-conscious mind sees things as black and white; it does not 'get' sarcasm, 'teasing' or 'jokes', it is a very literal part of our mind and we need to treat it with care.

It is a fact that has been manipulated by power hungry homo-sapiens since the dawn of speech. Brainwashing and mind-control work on the sub-conscious mind with often devastating results.

Fortunately it can also work to help us and it is never too late to start saying 'I like to be on-time'; I like being tidy; I like having a smart brain; I like to Eat Well, I like being slim'.

Picture the scene...

Ellie is sitting on her bed-sofa, staring at the T.V. flicking through the channels. She has a bowl of popcorn in her lap and is mechanically feeding herself.

She sighs and looks about her. She had so looked forward to being independent but never imagined this. Everything in her bed-sit is a bit shabby and miserable; the carpet is thread bare; it's dark outside, the curtains are drawn; the light is dim.

Suddenly a piece of popcorn sticks in her throat, Ellie grabs a can of coke and washes it down, spilling some down her front.

Swearing she gets up to put on a clean top and catches her reflection in a mirror.

'God, I'm putting on weight.' she mutters.

Changing her top she turns round and sees the popcorn and can of coke and has an epiphany moment ... suddenly she knows what she must to.

She pours the coke down the sink, puts the popcorn in the bin then fills two carrier bags with food from her cupboard –

biscuits, crisps, chocolate spread, packets of processed food, fizzy drinks ...

'I can do this' she tells her self, forcing a smile. 'I like being slim; I love being healthy, Suzi has shown me the way. I can do this; I have total control.'

Five minutes later Ellie is walking round the corner to a local supermarket. 'I can do this' she keeps telling herself. 'I am enjoying this experience.'

And suddenly she is.

She puts the food in a FoodShare box, takes a trolley and fills it with apples, tangerines, cucumber, a chunk of her favourite cheese, a bag of nuts and raisins; sugar-free muesli, honey, natural yoghurt, lean mince; a jar of Bolognese sauce; oat biscuits, lean ham and an uncut loaf from the bakery section.

A box of tee-lights catches her eye and bouquets of flowers at half-price and she thinks 'Why not?'

At the check-out she buys a magazine on Nutrition & Healthy Living.

The flickering candles are on the coffee table next to the flowers - infusing Ellie's bed-sit with a warm glow and the scent of vanilla. Music for Relaxation plays from her I-phone.

She sits upright on the bed-sofa, pulls the coffee table closer and gazes into the candles. 'I can do this' she tells herself.

'Meditation is as natural as breathing.' She tells her self. 'My breathing is slow and deep ... slow and deep ...

Her eyes close, she is conscious of the warmth from the candles on her face; the rise and fall of her diaphragm; she wills her muscles to release across her shoulders and in her jaw ...

'I love being in control, I love eating healthy food ...

And then her mind becomes still and all Ellie is aware of

is her slow deep breathing and the music and candle warmth washing over her.

A small smile spreads over her features.

We can all of us have epiphany moments when suddenly the way forward seems obvious; it's acting on it and having the intention that needs mastering.

Do have a look at the list of therapies at the back of this book that may support and help Techniques of Intention, particularly if your relationship with food is spoiling your life.

It's pretty obvious by now that our emotions and our relationship with food are intimately connected. Our mood affects what we want to eat and what we eat can affect our mood. Another loop!

Meanwhile, here's one Golden Tip for Eating Well.

WEAN YOURSELF OFF SWEETNESS

You will be amazed how quickly your taste-buds adjust.
A word about taste-buds.

If you look on the internet you will find astonishing information and images of our five different kinds of taste-buds located on our tongue, cheeks, epiglottis etc. These amazing teeny-weeny sensory organs can detect sweet, savoury, bitter, salty and sour flavours.

Now then, once you start reducing your sugary foods the sweet sensitive taste buds will start to de-sensitise and this happens quite fast.

By the third day of not eating any foods with added sugar and limiting your fruit to citrus and sour berries, like raspberries and blueberries, then on the fourth day you eat a banana you will find it almost unbearably sweet.

Most of us can't reduce sugar that quickly and a sugar withdrawal headache doesn't make it any easier but this too passes.

It's like reducing any addictive substance, take nicotine, for some people it works by gradually reducing the daily quota, for others a total ban works best.

Apparently the recommended amount of added sugar to have in a day is 25 grams (this doesn't include the sugar in fruit or root vegetables).

25 grams of granulated sugar equals six rounded teaspoons. (Not heaped!)

So, if you don't mind number-crunching then it may be useful at the beginning to keep count by looking at food labels on tins, packing etc.

If you don't like tea or coffee without a little sugar then include that in your daily total, even reducing it to half or even a quarter of a teaspoon per cup will make a difference.

You will quickly learn what items not to buy – it will become a new life-changing habit and one day you will look back and think 'did I ever really buy all that rubbish?'!

(It's been said that the average westerner consumes twenty-two teaspoons of sugar a day – oh boy!)

While we let that little lot sink in, here are some useful tips and old wives' tales you may or may not know and some are repeated (no apologies)

1. Eat like a king at breakfast; a prince at lunch and a pauper at dinner.
2. Before going out to a party have a carton of natural Greek yoghurt – it lines the stomach and stops us diving into the canapés etc.

3. Chew your food well, the stomach doesn't have teeth and you will feel fuller sooner.
4. Eat with chop-sticks – seriously – it takes a lot longer.
5. Never eat when feeling anxious or stressed.
6. When you crave chocolate make yourself a chocolate drink – cocoa powder, sweetened with a little honey, mixed with hot milk and water. This can be chilled and drank cold.
7. For healthy snacks chose walnuts, tangerines, raw carrot sticks, cubes of hard cheese, berries etc.
8. Protein foods keep us full longer than carbohydrates.
9. To stave off a hunger pang have a drink, even water can achieve this.

Speaking of water ...
Our bodies and brain need a lot of hydration; we are after all 65% water and every cell of our body – be it blood, muscle, nerve or gut exists in a watery saline solution. The brain is as greedy with water as it is with oxygen, in fact it is a very jelly-like organ; it has no bones, no muscle and is very wet and slippery. It needs to be hydrated!

When we're dehydrated nothing works well, even our muscles and joints suffer. Water is the basic ingredient of all our body fluids from blood, plasma, mucous linings, synovial fluid (joints), saliva and cerebral spinal fluid and we should be drinking at least five glasses of water a day.

I see many people with muscular-skeletal problems and for the majority, as soon as they increase their water uptake a lot of their pain disappears.

Water is by far the best liquid for us, however, being

clever and inventive and craving different taste sensations, humans have for thousands of years created many kind of beverages.

Not all of them good for us.

Sugar-laden drinks invented since the middle of the last century are disastrous for our wellbeing, helping to make us obese, diabetic and to suffer mood swings.

Coffee, tea and chocolate drinks containing caffeine may also affect our cell hydration as well as act as stimulants.

Alcohol has its own problems, not only is it addictive but it causes water to be leached out of the brain and other organs which is one reason why hangover headaches are so excruciating and why, if too much is consumed for too many years the liver starts to turn into a lump of 'clay' and ceases to function. (Cirrhosis)

Even milk is not an ideal liquid. Strictly speaking, milk is a 'food' and intended by nature for mammals to feed their new-born infants. Many adults find giving up -ow's milk improves their general well-being.

A feeling of thirst is easier to ignore than a feeling of hunger and when very busy we sometimes ignore thirst and 'forget' to drink, sometimes mistaking it for hunger.

The secret is to have a drink as soon as you feel your mouth and tongue are dry. Dry skin, dry eyes, a tickly cough from 'nowhere' and constipation can all indicate your mucous membranes are drying-out.

We literally wilt if we go too long without water, feeling lethargic and tired and just like a plant we lose our 'shine' and our skin becomes muddy. Headaches, too, are often caused by brain dehydration.

Oh boy! Isn't there anything to look forward to? Well, now for the good news:

Homo spiritus and food

When we are happy with our weight we have more energy and feel fitter; we feel good in our clothes, our self-esteem and confidence improves; we don't mind letting our partner see us naked – woo hoo, and life is much more enjoyable.

ALSO – we can start to enjoy having treats.

A treat is obviously not something we have every hour on the hour or even more than once a day. It's something we can look forward to once in a while – that's why we call it a treat.

We can decide on our favourite treats – for my mother it's Madeira cake and sea-salt chocolate – and then include them in our life to enjoy, now and again, as a treat, if we feel like it.

Your Limbic System will get used to this and savour the pleasure so much more.

That's all good and well but we're back to that Limbic System again which seems to over-ride our will power. If we're miserable, fed-up and anxious and worried how can we ignore the desire for instant pleasure we get from food and sugary/alcoholic drinks?

The obvious answer is not to feel miserable, fed-up, anxious and worried.

Okaaay ...

Another remedy is to have a life that is fulfilled – like our horse-lady ... to have something in your life that is BIGGER than your appetite, something that makes you forget about eating until you're really hungry because you're so absorbed in what you're doing.

Okaaay ... where's that magic wand?

No magic wand and yet there's some **very** good news for Homo spiritus:

If we've started taking on-board about tummy-breathing when stressed; the mona-lisa smile when anxious and 'holding' saliva in your mouth when nervous we have already started moving towards a new way of being. Yes?

The very good news is that this has been happening on a totally holistic level.

Holistic means body, mind and spirit.

So while we've been working hard on our will-power to reverse those bio-emotional loops we have also been changing the essence of who we are as individuals.

You see, there is a knock-on effect to making these changes in how we react to stress, a knock-on effect that almost goes un-noticed.

By reversing the bio-emotional loops of stress, anger, fear and by combating tiredness and ill health through meditation, a better diet and better sleep our general attitude to life begins to change.

We begin to *feel* less negative emotions and a change happens in mind, body and spirit.

We begin to feel more hopeful and positive because we are more in control and see our way more clearly and everything begins to change …. even our taste buds can change and we sub-consciously begin to desire the things that will enhance this sense of well-being, such as:

- Drinking more water
- Going to bed earlier
- Feeling a desire to be in the countryside or a park with trees and greenness. Lakes or the shore.
- Not craving caffeine or alcohol so much

- Seeking new interests, new companions.
- Watching less TV.
- Spending less time on face book and computer games.
- Maybe looking for work or a career which gives more satisfaction.
- Needing a bit more 'me' time.

Now, by taking on-board Eat Well we may find you start listening to our instinct for what we 'fancy' as our bodies gives us clear messages when it requires certain nourishment.

We will be free from diet regimes; we will eat just what is right for us at any given time. Yes, it may be salt or sweetness or spicy; salads, meat, eggs or cheese or fruit, nuts ... and we will keep it simple and enjoy the taste, eating at leisure and focusing on the texture and the flavour.

Don't be surprised if 'sugar' starts to make you feel ill - as in a large slice of carrot cake. It's a sign that something has shifted – something for the better.

And those of us who love to cook will invent new dishes, new recipes or explore new ways of eating for optimum health.

Exciting times!

On top of all that, our appetites may have started to reduce and we will not eat nearly so often or so much and all because we are controlling the stress response and it is the stress response that makes us reach for the doughnut – the milky coffee.

Our stomachs will start to shrink as we eat less.

Our digestive processes will improve, too.

The Loop of Over Eating has already started to Reverse!

And guess what? That mountain suddenly isn't nearly as hard to climb, in fact it gets easier the higher we climb ... not harder and a slip-stream is created so that others, coming behind us, find it easier and reach the top almost without trying and then everything changes.

A Quantum Shift happens.

Okay, now's the time to give the jelly a gentle poke ...

Put simply:

Quantum Physics tells us that everything is energy

Therefore, at a Holistic Level everything is Energy.

Mind, Body and Spirit is Energy.

Put even more simpler:

If we don't eat, drink and sleep we soon feel weak and lack energy.

Our thoughts and emotions, if excessive, can exhaust and drain us.

Our emotions, our thoughts and our beliefs and the food we put in our bodies all add-up to the Essence of who and what we are as individuals and then as a collective whole.

Our Intentions (both good and bad) are as real as our laughter, our tears, our muscles and bones.

Our pain and joy, suffering and contentment are as real as the chocolate we eat or the wine we drink.

At an energetic level nothing is separate, everything is connected.

All the loops we've looked at so far, illustrate this.

Humans, however, have lost sense of this connectedness.

What happened is, due to our very strong sense of Self and our freedom to choose, we have for thousands of years believed we are separate from everything; from one another, from nature, the earth and the sun, elements, atoms ... the universe.

We are too much in our own heads; too tied-up with our emotions – with how we are *feeling* and so jealousy, fear, hatred and greed took seed in the human genome as much as a tendency towards kindness and empathy.

To compensate for this bewildering (often terrifying) existence we compensate our distress with the pleasure of food, beauty, wealth and fame.

Nothing wrong with this, except ....

We haven't mastered owning Freewill; we keep getting it wrong, making destructive choices, over and over.

We haven't mastered knowledge of Self, a kind of 'insanity' surely started creeping into the minds of humans as they evolved a sense of Identity. The ego can be a terrifying experience, while for others it is an altar for self-worship.

Not good.

So ... we compensated for our failings by trying experiments using Power Structures as in political, social, religious doctrines, even medicine – in an effort to force people to live in a more 'sane' way.

Often brave experiments with good intentions but the way to hell has been paved over and over with these good intentions.

Those of us not in Power resist being forced, so Power has to punish, imprison, persecute even destroy the very society they try to control.

It is still happening, it will always happen until and unless our fundamental nature changes.

But that kind of change cannot be forced.

We cannot push the river.

Controlling our basic desires does not succeed if it is external. Change should happen from within each of us, particularly those in Power who should lead by example.

The thing is ... our failure to live peacefully is a tragedy over and over. It takes eons for a life form to develop a sense of Self and Identity. It is a rare and precious 'gift' and may never happen again on this planet.

It is a huge step forward into the Realms of Consciousness which have no boundaries of any kind. Not in Time or Space.

Home spiritus is very aware of this and understands how every thought, word and deed impacts the realm of Consciousness. Homo spiritus take great care of what they think, say and do.

The change to Homo spiritus is a quiet, inner revolution. Nothing about being Homo spiritus can be forced. It is not about abstinence or denial or following a creed or rules, or about reward or punishment.

It is about each individual – when (if) each of us is ready to take those steps the decision will be ours and no one else's.

A little awareness and knowledge is all it takes to make that first step.

Okay, deep breath ...

We've had a peep at the jelly and it hasn't collapsed, has it? It's still too soon to turn it out but beneath the mould its shape is taking form.

So, we are now ready for the next step because making choices is what being Human is all about but there's more to life is than breathing, sleeping and eating and once childhood has passed it's all about work and responsibilities – there's just no getting away from it.

That means STRESS. It means worrying about money, jobs and relationships; about becoming independent.

Wouldn't it be great if we could limit stress in the first place?

Home spiritus can.

Okay, take a deep breath ...

STEP 5

LIVE WELL

Okay, so we're learning how to gain a healthier life; we know how to reverse our bio-emotional reaction to every-day stress; how to get a better night's sleep and how to eat well.

We're beginning to understand that the choices we make have far reaching consequences, that our one life is very precious and to have self-determination is a wonderful thing.

But it's also tough; it's like being alone on a small boat on a wild sea and then discovering there's a rudder we can take hold of and steer the boat into calmer waters, a safe harbour or port, maybe.

The question is: Now that we're no longer children and sailing in our parents' boat where do we want to sail to?

How can we make our journeys safer and more pleasurable?

It's not an easy journey because an independent life involves money, time and work.

Did you groan just then? Did you slump a little at the very mention of money, time and work?

It doesn't have to be that way.

Okay, the sea exists; it is there whether *we* exist or not. But we get to choose the type of boat, and the rudder in our hands can guide us to wherever we choose, though we have to work at it.

It always comes down to choosing – that is both the curse and the joy of being human.

We sometimes forget we have the power to choose. We can so easily allow ourselves to get blown off course or allow someone else to hijack our boat and take us where they think we should be going.

If we know how to live well, we can make better choices, wiser decisions and become the person we were born to be.

Homo spiritus is that person, humans have existed for enough millennium, made enough mistakes, caused enough misery, to finally 'get it'.

Get what? Well, the truth is just around the corner, a truth that is at first painful but then rewarding and can eventually 'free' us from our ancestral chains.

Be patient, the jelly is setting.

Let's first look at money, that stuff we've invented to measure the value of something, be it an object or another person's labour. Clever idea ... very clever of old Homo sapiens, however ....

Money ... money ... money

'Must be funny, in the rich man's world.' *ABBA*

It has been said that money is the root of all evil although greed is more likely to be the culprit because if used wisely and with respect, money has an astonishing 'energy', nothing short of spiritual. We just haven't realised this ...

Picture the scene...

A busy day at the shopping mall with shoppers and window shoppers; coffee-drinkers; a student busking by the doorway; a lady standing by the elevator, rattling a charity tin; a security guard on patrol.

A young woman, with purple streaks in her hair, long skirt, boots, denim jacket, exits a wall-paper shop with a roll of paper under one arm and a wicker basket over the other.

Being a Homo spiritus her eyes see everything; they see, in the reflection of a shop window, the boy, about sixteen with hunched shoulders, uncut hair, old sweatshirt and pants - walking up behind her, head down, hands tucked inside his sleeves.

She checks her breathing. She is calm.

He comes up close behind her; she remains calm. She is ready.

The shift of weight in her basket is slight but it is enough.

The boy turns on his heel but Homo spiritus is quicker, she turns also and in one smooth movement brings the roll of wallpaper down onto the boy's shoulder.

He yelps and stumbles, her purse dropping from his hand. His balance is poor; he falls to the ground and she drops with him, picking up her purse on the way and kneels on his shoulder.

'Listen well,' she whispers. 'And don't get up'.

Her eyes shine with fierce light, the boy feels unable to move, both shocked and mesmerised.

From the corner of her eye she sees the charity lady gasp and the busker stop playing; people stop and stare.

She leans in and whispers in his ear. 'If you want money, sing for it; ask for it; work for it but don't steal it.'

'Everything all right?' Someone asks.

Homo spiritus holds up her hand – keep back. The people freeze. Her eyes stay fixed on the boy.

'My name is Suzi and I have a message for you.'

Opening a side zip in her purse she takes out a small wad of money and presses one note at a time into his hand.

'First you pay what you owe.

Then you buy what you need.

Then you look at what you have left, if anything.

One half of what is left is yours to do what you wish – save it, spend it.

The remainder you give away – as a gift, expecting nothing in return.'

She closes his fingers around the money.

The security guard is running down the escalator, talking into his mouth-piece - a shop assistance has called-in a disturbance on the ground floor.

'Do you understand?' She whispers.

The boy nods, speechless. She hauls him to his feet.

'The magic is this ... what you give away comes back to you ten-fold. You will never go hungry.'

She embraces the boy and whispers in his ear. 'Teach this message to one other thief and your debt to me will be repaid. As mine is now re-paid.'

Smiling she gives him a wink then turns and walks towards the elevator. The security guards stands and scratches his head, people move on.

Homo spiritus pauses and turns to watch; the boy is looking at the notes in his hand, she knows his struggle and she wills him to do it. Slowly he walks toward the charity lady; he hesitates then stuffs a note into the tin.

The charity lady beams – 'thank you!'

The boy nods and walks away, his shoulders relax; he lets out his breath and lifts his chin.

In 2017 the world is changing fast – it is scary, uncertain, confusing. Society is changing; Banks, Insurance, Investments, Stocks & Shares, Pensions, Savings, all a whirlwind of doubt, mistrust and corruption, and Debt – both national and on a personal level is spiralling out of control.

Money – a means of valuing labour or the worth of

something beautiful or useful, is now a commodity in its own right and its energy has changed from something that should flow from hand to hand to something that is horded, stagnant or exists as strings of valueless numbers, a vapour with no substance.

Corruption, greed, booms and bust, recessions and inflation rock the core of our societies and it will take a new kind of Human to figure out how to rein-in the beast so that it serves us rather than consume us.

Homo spiritus will arise from all sectors of society in an organic and seamless way. They will be financers, as well as inventors, teachers, medics, lawyers, engineers, retailers, entertainers, entrepreneurs, agriculturists, politicians – even.

A few already exist and their work is quiet and unnoticed, by most. This is the way of Homo spiritus – trumpets are not blown, stages are not erected, interviews are rare – the ego of Homo spiritus is strong but gentle.

As more and more evolve and when all the dots have joined and we have conquered our fears then the answers to all the ills of our societies will also evolve as natural as day follows night, until then the 'parable' of Suzi with a message for the boy-thief is a good place to start.

A gentle start.

Coercion, threats, bullying, bribery and blackmail are the old Homo Sapiens way – and change won't all happen at once – it will not happen overnight, it will happen over time – just one company, one business owner, one town, one individual ... and then another, all quietly becoming the change they wish to see in the Janus* world of money and finance.

(*Look-up this word, if you like.)

Results will speak for themselves until finally a Critical

Mass is reached – a huge Quantum Shift occurs in the consciousness of the human mind.

We will think differently, our mindset will change and Homo spiritus will look back, as we look back today at the things our ancestors did in the name of ignorance, fear and superstition and Homo spiritus will shake *their* heads in wonder and disbelief.

It is time to change, time to face a new direction.

Now then, as far as financial Loops are concerned it's pretty obvious we get into debt because we over-spend (just as we over-eat) and then we are forced to spend money we haven't yet earned. We are often persuaded to borrow more money to pay off our debts. Loop-de-loop-de-loop.

Why do we do this?

Some of us can remember when consumerism was a new word.

We can remember a time before supermarkets, hypermarkets, Malls and on-line shopping.

We can remember when shops were the bakers, butchers, fishmongers, a hardware-store, a sweet-shop and a general store.

Shoe shops, dress shops and gentlemen-fitters were found in the larger towns or cities as were toy-shops.

We remember our mothers sewing our clothes and teaching us to knit. We had one winter coat and two pairs of shoes. Gift shops were rare, found at sea-side resorts to buy souvenirs of our annual holiday. Toys were limited to birthdays and Christmas and when Grandma and Grandpa came to visit.

In short, we weren't tempted to buy an excess of consumables or knick-knacks simply because they didn't exist in the first place and yet it was always exciting to have a new coat; a present from Grandma, sweets on Saturday.

It is that excitement of having a treat that triggers the Limbic System – the Pleasure Response. Remember?

Back in the day our expectations were small but that all changed with Consumerism and the face of Janus turned to a new direction and we became addicted to shopping and the pleasure it gives us. (Just as we do with food).

Advertising knows how to capitalise on all this, especially with the advent of television and glossy magazines, the Internet and on-line shopping. So now we are seduced into thinking that a new car, a new gadget, an exotic holiday, all these lovely rewards for all our hard work and stressful lives, is ours for the wanting.

Credit cards were invented to help us 'pay' for things we can't afford, followed by store cards with enticing offers so it is even easier to get into debt.

Spending money we have not yet earned has become the norm ... in fact we are considered quaint or unworldly if we don't have a credit card or two ... or three.

The problem is this - getting into debt creates more anxiety and so we need more treats to make us feel better!

Whoops, another devastating loop.

Can we reverse it?

The obvious reversal is to deny ourselves those treats; live a frugal and minimalistic life; cut-up the credit and store cards and get out of debt. This can take years, enormous will-power and for most Homo sapiens it would mean a miserable existence.

Now, this bit's important –

Once we start reducing our bio-emotional response to stress and improve our eating and sleeping, our 'behaviour' changes, the stress-habits we have fallen into begin to shift.

We no longer feel the need to impress. No longer feel impressed by a celebrity life-style, celebrities too will feel this change; some are already feeling it.

Our need for possessions diminishes yet we will still adore beautiful things. The world of Homo spiritus is not a grey world of austerity, it actually blossoms into something far more beautiful.

The thing is this, envy ... like greed, fear, anger and hatred will slowly fade from the human psych as more and more people move further away from ancestral patterns of behaviour.

Personalities too will blossom and each will find their own sense of value in what they do for a living, be it creative, artistic, practical, of service to others, in sport or entertainment. We will find a new sense of value in our relationships and where we chose to live, be it a city, town or village.

The future is not about making everyone the same – far from it; it is not about denial or deprivation – far from it; it is about finding ourselves and realising our potential, if that is what we wish to do.

Even more importantly we will value helping one another and we will celebrate everyone's achievements, small or large.

Beautiful cars, works of art, jewellery, clothes etc will continue to be designed and appreciated by Homo spiritus and available to those who wish to own them.

All of this may sound like an impossible dream because this is not how Homo sapiens behave; we are often envious of people who are successful - who drive expensive cars or have a lavish life style; who win competitions or the lottery.

We have an ingrained sense of 'unfairness', not always appreciating that some life-styles have been well earned by sacrifice, long hours, risk and dedication.

However, it is also true that some individuals and societies have 'it all' and in excess, while others have literally' nothing' and go hungry, depending on world charities for subsistence.

Something is wrong.

The divide between rich and poor has never been so obvious and there is a growing awareness that World Economies are on the verge of collapse due to mismanagement, corruption, tax evasion and other criminal activities.

Something is very wrong.

As one more person, one more consumer, one more celebrity, one more employer or employee, or one more politician begins to reverse their Homo sapiens bio-emotional loops and discover what is really important then what is wrong can be put right.

Society will change and the way it is governed will change.

Money is going to change.

Advertising will change.

Capitalism will change.

Banking will change.

Economics will take on a whole new meaning.

It's hard to imagine how all this can happen if we look at the immense wealth, control and influence that some industries have, such as weapon manufacturers and pharmaceuticals.

Homo sapiens believe that we need these industries for our safety and wellbeing.

Just let that last sentence sink-in and see if you can spot the irony and then think what would happen if people stopped buying guns; if people no longer needed medication for diabetes or high blood pressure?

The logical conclusion is that if no-one is buying then no-one is making – supply and demand - the foundation of all economic models.

It takes only one person to say, 'hey, I don't need a gun': one country to say, 'we don't allow the manufacturing of weapons'; one Bank to say 'we don't finance the arms trade', to get the ball rolling.

If we can imagine that kind of scenario and believe in its possibility then we are already entering a new mind-set; the mind-set of Homo spiritus.

And here's another thought – all those brilliant brains, those clever men and women – the scientists who design and build the incredible war machines and the chemical formulae imagine what they could achieve in other undreamt of industries

It's not an easy thing to imagine, it just seems too far-fetched – just as our great grandparents could never have imagined PayPal, Twitter, mobile phones, the Internet, Amazon ...

And yet, something rather weird starts to happen as take back control of our spending, as it does with our breathing; our sleeping and our eating ... just as it becomes *normal* and not at all forced to start going to bed earlier and getting up earlier; to lose the taste for sugar, so too does it become normal to buy only what we need.

Today, money makes the world go round more than ever before; it is the engine that drives our existence and shackles us to life-long debt.

That can change.

Meanwhile, here are some inspiring tips concerning money, some of it random, picked-up along life's highway:

- 'Neither a borrower nor a lender be'. (Shakespeare)
- Look after the pennies and the pounds will look after themselves.
- Live within your means. If you haven't got it ... don't spend it.
- Budget your earnings. (Read thief's 'message' again)
- Have a savings-pot for a 'rainy day'.
- Put one penny in a savings jar, then the next day 2 and the next day three, and so on ... by the end of a year you'll have a nest egg for a holiday or celebration!
- Don't borrow to speculate.
- Pay your way – a free-rider has few friends.
- Don't buy cheap carpet – a good carpet will make even shabby furniture look good!
- Look after your teeth, your feet, your health – money spent on health is never wasted.
- Always thank someone for a gift, always.
- If you haven't worn something for more than two years take it to a charity shop.
- Do you really need another pair of shoes, trainers, boots?
- Be aware of how clever and seductive advertising can be.

- Don't expect something for nothing, if you do it will have no 'value' for you.
- There is no such thing as a' free dinner'.
- Be charitable if you can afford it.

If you're a shopaholic and have more clothes, shoes, books, make-up, gadgets than you have time or space for, then possibly this is a Compensatory Behaviour and may be worth investigating.

However, don't worry too much because as you practice steps 1, 2 & 3 the chances are you have already began to change this pattern of behaviour, and as you progress through steps 4,5 & 6 the loop continues to reverse – in more ways than you'd ever imagine – be patient.

This is the joy of our re-evolution – it is a Holistic Event, without the need for laws or dogma or rules. It will happen gently yet powerfully and Homo spiritus will look back and ask themselves – 'when did it happen?'

There will be no 'July 4th', no' Bastille Day', no '1066' , no D-Day ... nothing like that ... just a quiet and gradual change.

When humanity changes its mind-set then society changes too, it is inevitable. And if we succeed in changing how we react to our emotions of fear, greed and anger then the world will become a much fairer place.

Once we have a fair world we will have a world at peace. But it can't be forced or decreed by Laws. It can only happen by the conscious will of individuals and this can only happen when we stop being afraid that someone else has more than we have.

To do this we need to really understand what it is to

'share' – be it food, water, oil …

This can start to happen by practicing the steps explained in this book.

Individuals will start to experience more 'light bulb' moments and Inspiration will guide us to new concepts, new methods, new inventions, new music, art and science.

It will all take time, maybe several generations but that's okay. A house is not built with one brick.

So, a change in our attitude towards money is inevitable if we take these steps forward, so what's next?
What else is stressful for Homo sapiens?

WORK

'If you find a job you love, you'll never work again'.
Winston Churchill.

Picture the scene...

Halls of Residence at a University … a landing with doors leading into bedrooms, some open, others closed.

Sam bursts through a half-open door … oops, wrong door … (or is it?)

'Err, sorry mate!' Sam swings on the door, putting on the breaks.

Homo spiritus is at his desk; he turns and grins, a flash of white teeth against his brown face. 'No worries.'

'Hey, cool room!' says Sam.

He looks around at the polished wood, the maps and posters on the walls, African artwork, woven rug on the floor, African-sunset throw over the bed and clothes hanging on a rail. Shelves are crowded with books, framed photos of elephants, people and gleaming sport trophies.

'What's your sport, dude?' He asks.

'Basket-ball.'

'Cool. Wish I had a room like this.'

'You do.'

'Huh?'

The Homo spiritus got up from the desk and smiled. 'Sure you do, I've seen you … next landing down.'

Sam blinks. 'Hey, dude, my room is not like this.'

'What's your name?'

'Sam.'

'Hi Sam, I'm Luther, c'mon I'll prove it to you …

'Wait! Where are you going … stop …'

Luther bounds down the stairs, Sam in his wake, he sees the room with a sheet of A4 - 'Sam's pit' on the door – nice bit of graphics – pushes the door open and steps inside.

Sam follows, embarrassed, pushing his fingers through his hair.

Luther steps over the clothes, empty food containers and beer cans and walks to the window; ignoring the smell, the mess and the un-made bed. 'See, same window, same desk, lamp, shelves, bed and … through that door is a shower and toilet, right?'

'Don't go in there!' Sam yells.

Luther smiles and walking to the shelves picks up a tarnished sports trophy. 'So, what's this cup for, Sam?

'Orienteering.'

Luther nods his head. 'Follow me.'

Taking the cup with him Luther retraces his steps.

'Wait!' says Sam.

Back in his room, Luther takes a cardboard box from under his bed and pulls out a tin and some wadding and a soft cloth.

'There you go, Sam. That cup deserves to shine.' He puts them on the bed. 'Fancy a brew?'

Sam sits on the bed and picks up the tin.

'Instructions are on the back.' says Luther.

Sam polishes his trophy. The room is quiet. He is absorbed.

Luther comes back from the kitchen with two mugs of tea. 'So, what degree are you doing, Sam?'

Sam grunts. 'Social sciences'.

'Uh huh.'

'You?'

'African studies.'

Sam looks up and smiles. 'Roots and all that?'

'Maybe, I just fell in-love with elephants when I was four.'

Sam grins.

'So, how's your course, going?' Luther asks.

Not what I thought.' Sam sighs. 'Mistake to be honest, bad choice.'

'It's still first semester, you've time to change.'

'Yeah, but what to?'

'When you were a kid ... what was your passion?'

'Huh?'

Playing ... what did you play at?'

'Uh ...'

'Before you got into computer gaming ...'

Sam laughs. 'Camping, I loved camping, always making tents, wigwams and hides, in the garden – you know out of junk then pestering mum to let me sleep outside.'

Luther studies him for a moment. 'Survival programs? Guess you like ...

'Yeah, man, stuff like that – awesome.'

'So ... there's your answer. Go for it.'

145

'How?'

Luther sits down beside him.

'You'll find a way.' He points to Sam's forehead and then the centre of his chest. 'In here, and here ... you'll find a way- hold on to the dream, believe it ... see it.' Luther sweeps his hand, taking in his own room, his own dream. 'Talk about it, bring it into your reality. Put up posters, maps. Doors will open – it's what happens.'

Sam stares at the Homo spiritus.

'Is that what you do?' he asks

'Yep. Get rid of the clutter, Sam. Get rid of the mess - keep everything shining and you will shine. Trust me.'

Once we shine, shadows caused by clutter in our mind and under our bed, disappear as our intuition lights our way.

Okay, so shining takes some effort and many of us find effort a chore and chores and an effort. Why? It only takes a few minutes to do a task – but we procrastinate. Humans put off doing chores on a daily basis. Why?

The truth is, it's hard work living on a rocky planet governed by gravity – dust, weeds, dirt, rain and that's on top of working hard to put food on our tables and a roof over our heads and to keep safe and well.

It's daunting, relentless and exhausting, that's why we need holidays to rest and to 'step off the planet'.

That's why we learn skills and get an education so we can find work so we can feed, clothe and house ourselves.

Phew, it's tiring just thinking about it.

So what about Homo spiritus?

Before we answer that one let's look at another list of tips and advice that may help us keep a track of time and our work less stressful.

1. Before you've got dressed and gone downstairs in the morning, make the bed, tidy your bedroom, clean the bathroom/toilet – you won't go upstairs again until evening when you're too tired to bother.
2. A stitch in-time saves nine. Put off dong something and the job just gets BIGGER.
3. Tomorrow never comes, do it NOW.
4. House-elves don't exist (sob).
5. Make a list of 'things to do' to help keep on-track.
6. To help find a job or career that will suit you, take a few of the many psychological career-tests on-line. They're fun to do and illuminating!
7. It's never too late to learn a skill or to study for a qualification.
8. When you're busy time is short so allow plenty of time to get to work, to an appointment – Homo spiritus does not rush. Homo spiritus has mastered time.

Speaking of which …

Picture the scene...

It's Friday evening and Sam and his Uni pals have gone to eat at the Mexican Diner. It's noisy and colourful and Sam recognises the waiter who comes to serve them their drinks.
   'Luther, my man!'
   'Hi Sam, how ya doing?'
   'Cool, hey everyone this is Luther, he's in my Hall.'
   'I know you,' says one of the lads. 'You're captain of the

basketball team, aren't you? I've seen you play.'

'That's me.' Luther flashes everyone a grin and puts down their drinks.

'Yeah, you got your nose broke, I as there. Coach fixed it, didn't it hurt, man?'

'Not really.'

'Didn't know you worked here, dude?' Sam says.

'Started on Monday – tips are good in this place.' He winks and they all laugh.

'How'd you do it, man?' Sam asks, shaking his head. 'Don't you sleep?'

Luther laughs and looking over his shoulder sees his boss, she glances his way and Luther holds up five fingers, she nods and flashes him five in reply.

Pulling out a spare chair Luther sits astride it, cow-boy fashion. 'Boss has just given me five …'

The others glance round, had they missed something?

'Cool boss.' Someone remarks.

Luther nodded. 'Yep. So, Sam, how do I do it? Well, I'll tell you.'

Leaning forward Luther drops his voice and they all bent forward to listen.

'It's all down to number crunching. My boss has given me five minutes but she knows she'll get it back. Time, dude, is like money. You can spend it wisely or you can waste it. Am I right?'

'I guess,' said Sam.

The others nodded.

'Now you gotta admit that twenty four is a pretty awesome number.'

They all blink. Faces blank.

'C'mon guys, twenty four hours in a day? You can crunch that number so many ways – eight hours sleep, eight hours *work, eight hours leisure. Or ...*

'Yeah, twenty four has eight devisable numbers,' said one of the girls. 'And it's the smallest number that does.'

The others roll their eyes. 'Kelly's our geek,' said one.

Luther laughed. 'She's right, you can mix n match those devisable numbers any way you like. The secret is this ... first you spend your time doing what's got to be done – sleep, eat ...'

They all cheer.

'Then what has to be done – cleaning, studying, working ...'

'They boo'.

Luther grins. 'Then, whatever time is left, if you're clever, you do whatever you like.'

The table cheers.

'It's also cool to give another person some of your time and that's always possible if you've time to spare. Well, I've had my five ..., catch you later, Sam. Enjoy your meal, everyone.'

Standing up Luther swings round the chair to face the table again, in one smooth movement. 'Oh, and another thing - in basketball the offensive team has twenty four seconds to attempt a shot.'

He chuckled and walks away.

'He's one cool dude,' someone says.

Kelly sighed. 'Damn cool.'

The thing about work and time is to feel in control of both, as with money.

This is not easy and is part of learning how to be an independent adult and it starts when we reach senior school ... as though puberty isn't enough! But there it is, that's life and it can be a good life -if you've got your hand on the rudder.

So, whenever school, your job or your time (lack of it) starts to make you feel panicky or stressed this is what you do:

1. First - Reverse the stress loops – take a few seconds to breathe, drop your shoulders, unclench your teeth, smile.
2. Check your time-management, were you late or rushing or putting-off. (Ooops.)
3. Are you happy what you're doing?
4. Are your worries, relationships, effecting your concentration?
5. Who can you talk to about this?
6. Understand you're not alone, people do care but no-one can read your mind.
7. Check you're not being swayed by other's people influence, if something doesn't feel right for you, move away from that influence or speak your mind – 'actually Mum I don't want to go to Uni, not yet anyway'.
8. Communicate. Let people know what *you* think is best for you, listen to any advice then think about it.
9. Avoid knee-jerk reactions. Once you're in control of your emotions then speak to that boss, friend, parent, sibling, school or work colleague why they've made you upset or anxious, troubled or unhappy.
10. Be prepared to change course – life is journey of discovery, avoid feeling trapped. It may take several jobs and disappointments before you find your life's purpose however humble that may be.

11. Plan that change before jumping in with both feet. (No I'm not going to Uni but don't worry Mum, I've found a job at the dog grooming parlour.)
12. Even if others think you're crazy, follow your dreams (having prepared the way forward) and just ask for their moral support.
13. You'll be amazed how the world will support a Homo spiritus.

Okay, maybe you're starting to feel overwhelmed by all this hard work – chores, jobs, decisions, and inevitable stress and worry about money, health, relationships.

Even reversing bio-emotional loops is hard work!

Happily, Humans discovered a knack of handling stress by doing something quite magical.

We play.

Humans love to play.

Humans at Play

You only have to visit a sea-side resort to notice how clever and inventive we are at playing; ball games on the beach and building sand castles, arcades, putting, crazy golf, kite flying, water sports ...

Humans need to play, we need to entertain and be entertained and it is astonishing the lengths we go to; the toys, games, pastimes and gadgets we have invented to amuse ourselves.

We put on plays; musicals; puppet shows; pop-festivals; we go camping, cycling and have invented all manner of games and sports using a spherical object from marbles to basket-ball; then we included a 'stick' and we have golf, tennis, base-ball ...wow!

And then we have the many board-games using numbers and mental challenges, precision, strategy - such as chess and card-games.

To be inventive with playtime is a gift we have from early childhood and as we get older and reach adulthood the best remedy for stress is pleasure. (Remember the Limbic System?).

It's obvious why, for when we are entertained or when we 'play', we smile; we laugh; we focus and we forget our worries for a while and the stress hormones switch off.

To truly laugh and enjoy ourselves is powerful medicine and crucial to our wellbeing and happiness and is a key to our future existence.

Having such a strong sense of Self we are consciously aware of feeling joy and happiness, we actually glow and the feeling is infectious. When someone is belly laughing – a toddler in a high chair or two old ladies on a bench we can't help but catch-on to this feeling.

The ability for humans to feel joy transcends us, it lifts our spirits.

We are still transcending.

The point is this, playtime is essential to our continued evolution, it always has been;

it is an expression of our imagination which plays a huge part in our evolution. Even Einstein said Imagination was our greatest gift.

Wouldn't it be cool to know when, in our long evolution, humans started playing with sticks and stones during time of idleness?

Who knows, maybe Stonehenge was a sport stadium? (Only kidding).

Sport is big business for modern humans, from the

Roman gladiators to the spectator sports we have today and always will be, but cheating and corruption, so prevalent today will become a thing of the past.

Competition will not be about money or ego but a thrill in each and everyone's accomplishments.

For Homo spiritus playtime starts to become a bigger part of our existence. We will have more time to play, we will be less tired and as our love of life and sense of joy develops so does our evolution.

Sounds unbelievable?

This is how it works:

By controlling our bio-emotional response to stress and by recognising the things that make us stressed, i.e. poor time management, poor money management; compensatory behaviour patterns, over time we realise a new way of being creeps into our leisure time too.

We get to be really good at it!

Let's face it, a lot of our games and pastimes require a certain level of focus or skill, maybe a steady hand or a good eye. Think of ... darts, snooker, golf, houses of cards, model making, arts and crafts.

Patience and a good memory may be involved, think of card games, chess and other board games.

The magic is this – by breathing correctly, relaxing shoulders and jaw muscles; by pulling saliva into your mouth and smiling that tiny smile to yourself as your swing that putter, place that card, take that shot - your control is absolute and thus your aptitude reaches new heights.

Failure or the ego-centred desire to win, at all costs, don't matter nearly so much, there's always another game and 'well done the winner!'

To play with others who are better than ourselves becomes a privilege as it will help us improve and also the reverse, those who are champions will honour helping other to achieve their potential.

Trust me.

Here's a true story...

Many years ago I had my first and only experience of clay-pigeon shooting. Always ready to try something new I agreed to have a go. So ... I followed instructions, watched the others and then it was my turn.

Taking my stand as instructed and to combat my nerves and hoping not to make an idiot of myself, I immediately went into correct breathing, saliva etc, grounded my feet firmly on the earth, shouted 'pull' raised the gun, saw the clay-pigeon arc in the sky, followed it with the tip of the gun and squeezed the trigger. Bam!

I was totally surprised to see it shatter into small pieces – it was so easy! The instructor and everyone else there that morning were impressed. 'Beginners luck' they said.

About fifteen clay-pigeons later – all different angles and even some doubles - bam bam! They had changed their minds, as each one was a direct hit. In fact I seem to remember a deathly silence.

The instructor, though, was incredibly excited and wanted me to join his shooting club but my shoulder was bruised and it really wasn't my thing but the point was - had there been time to explain how I 'did' it they could have been just as successful.

So, using the powerful combination of stress-management plus better time and money management any natural skill or sporting ability can be fully developed.

This is the future of Homo spiritus and the same is true

of the non-physical pastimes such as music, theatre, film, Art.

Human creativity knows no bounds and we have much more to invent, create and imagine and Homo-spiritus will have more free-time plus freedom from anxiety to focus on and achieve their dreams.

Bring it on!

So, is that it?

Have we taken all the steps needed to start a new journey towards a more evolved Conscious Being?

Is there anything else apart from our stress-management, health, wealth, work and leisure to examine?

Oh my lord, yes!

Step 6 is the Biggy – it is HUMONGOUS!

The next and the last step is the one that has always tripped us up and yet it is also the one that will lead us away from our self-destruction.

It is also the hardest to take.

Why?

Because for most of us it's more like a leap over the abyss than a step in the sand.

Because, this is when we turn out the jelly and, maybe, we don't like its flavour.

Because this is when Copernicus announced that it isn't the sun that moves in the sky, it's the Earth going round the sun.

This is when we learn something that goes against our natures.

To be honest, what we are about to hear we've heard before, it just seems that Homo sapiens never took it fully on board.

We think we have, but we haven't; not really.

If we had we'd not be where we are now, laying waste to

a beautiful planet and fighting endless wars.

The thing is, this leap is totally **alien** to our way of thinking and our way of behaving. It goes against all our bio-emotional responses.

For this reason we may feel a strong resistance, even anger ... read it once and say:

'No way!'

'You must be mad!'

'That is simply asking too much!'

If that is how we feel then let's read it again, breathe, smile, relax and with an open mind say 'okay, let's give it a try, we've nothing to lose ...'

STEP 6

LOVE WELL

'Love, Love, Love ... All you need is love'. The Beatles

In the 1960's The Beatles was a hugely successful pop-group, one smash-hit after another, alongside other bands such as the Rolling Stones, The Beach Boys ...

Each generation have their own taste in music and clothes. That's what puberty is about – young people expressing themselves, becoming unique, separating themselves from their parents' generation, becoming independent.

It can be painful, troubled, fraught with emotions (oh boy), it is also one way in which free societies move forward.

For Homo sapiens each generation has been a march forward into new ideas, new horizons. This has been an essential part of our evolution, otherwise we get stuck in a time warp and then we stagnate and stagnation is not health, it leads to death and decay.

The interesting thing is, when it comes to music it doesn't matter which generation or style of music - be it blues, soul, folk, pop, heavy metal, rap, hip-hop - the lyrics are massively concerned with 'love'.

Not just music but fiction, poetry, films, theatre – just about all our creative arts have at least some element of 'love' in its theme. Take the James Bond movies – about spies and danger and spills and thrills and yet there is always a love interest.

Why?

Because it 'sells'?

Certainly, but then, why does it sell?

Why does our hunger for love - to be loved - to find love, consume us?

As children we are generally loved by our parents and grandparents but this is no longer enough, in fact it is something we feel the need to break away from. Yes?

Okay, let's get one thing clear – there is familial love then there is sexual love. Remember we talked about sex in step 1?

It seems we are still wired to an ancestral need to find a mate. All animals are driven by this need; it is the fundamental urge of nature to reproduce.

We humans like to think it is different with us; we like to believe we transcend that basic urge and seek a mate who we can share something special with.

Some couples achieve this kind of bond but all too often it breaks down and when that happens, oh my! anyone who has experienced the pain of a break-up know the depths of emotional pain this can cause, not to mention litigation, divorce, and worse.

There are enough songs, films, books about human heart-ache to fill another Olympic stadium.

The problem is, we often expect too much from love, or too little; in fact, we're not totally sure we know what Love is …

Picture the scene...

A small lecture room at Uni – on the door is a notice that says-
  'What is Love?'
  Tonight 19.00
  All welcome.
  On the podium is a large white-board and a small table

with the lap-top computer. A song is playing, a classic - 'I want to know what Love is' a power ballad by the British-American rock band Foreigner.

The tutor checks her watch as students begin to file into the room. She encourages them to come down to the front rows. By two minutes to seven about thirty people, including Sam and Kelly have arrived.

One minute later the facilitator turns down the volume, waits for everyone to settle then introduces herself.

'Good evening everyone. My name is Jennifer, I'm a lecturer in Social Sciences and Life Skills and I've been coerced to lead this discussion on 'What is love?''

There is a slight chuckle.

Jennifer looks tired; she is tired; it's been a long day. She tugs her fingers through the ends of her wavy hair and suppresses a yawn.

'Some of you, I know, have also been coerced, by your tutors to attend this discussion as part of various modules so let's get the show on the road.'

Moving towards the white-board she picks up a black maker pen. 'We're going to write a list of words, from A to Z that describes what Love is, okay? Shouldn't be too difficult, but let's keep it clean, right?'

Another chuckle.

' So ... someone give me an 'A'.'

She waits, pen poised.

'Affection', says someone.

Jennifer nods. 'Okay'. She writes the word on the board. 'B' please, someone.'

A few seconds silence.

She glances at them, eyebrows raised.

'Beauty,' says another voice.

Jennifer sniffs. 'Okay'. She writes the word.

Then she writes the letter 'C?' and glances at the students.
'Compassion', someone says.
Jennifer nods. 'Good one.' She writes the word, 'Now, a D please.'
 Silence.
People fidget.
Jennifer sighs and is about to speak when a voice is heard.
 'I have a list.'

It is the soft voice of a Quiet Man, sitting to one side, not young, not old. Cargo pants, a checked shirt, rolled at the sleeves, unbuttoned at the neck. Thinning brown hair, glasses ...

Jennifer blinks, she doesn't recognise him.

Sam does and turns to Kelly and whispers. 'I know that man.'

The man stands up, he smiles, his eyes crinkle.
'Martin. Mature student. History. May I?'

He walks towards the podium, holding Jennifer's eyes with his own and gently takes the black marker pen from her fingers. 'It won't take long.'

She blinks again and moves to one side.

'Love is Anger' he says, softly but loud enough for everyone to hear, and writes 'Anger' at the top of the board.

'Love is Betrayal'. He writes Betrayal under Anger.

Jennifer sucks in air. 'I don't think ...'

The Quiet Man turns to her with a finger on his lips, he smiles.

'C is for Crying.' He says, and adds this to his list.
 The room is quiet, very quiet and no one fidgets.

For several minutes he says and writes twenty six words. When he has finished the Quiet Man stands to one side and studies his list:

Anger

Betrayal
Crying
Deceit
Envy
Fear
Grief
Hate
Injustice
Judgement
Killing
Lying
Misery
Needy
Obsession
Possession
Questioning
Revenge
Suspicion
Tyranny
Unbearable
Vengeful
War
Xenophobia
Yelling
Zealousness

Jennifer is at a loss, she drags her fingers through the ends of her hair again; she has to say something. She turns to the students.

'Is Martin right? Who agrees that Love can be all these terrible ...'

Her voice trails.

No-one answers her. They all stare at the Quiet Man who

stands with head bowed and a small, sad smile on his lips.

'Yeah, I reckon he's right,' says a new voice.

It is Luther, sitting at the back, his long legs resting on the seat in-front. Heads turn, he holds up a hand in greeting.

Sam turns his head. 'Hey, Luther.'

Kelly turns also and waves to Luther who waves back.

So ... sir, can you introduce yourself and explain why you think Martin is right?' Jennifer asks.

Luther stands up; smiles and shoves his hands in his pockets.

'Sure; I'm Luther, first year, African Studies and Anthropology and I think Martin's list is right because Love has a dark side.'

'How do you mean?' Jennifer asks.

'Well, if you love your country you can become a patriot who becomes xenophobic. If you love your God you can become a Zealot who kills in the name of that love.'

There is a collective intake of breath.

Martin lifts his head and grins at Luther; one Homo-spiritus, recognising another.

'If you love someone with a fierce passion you can become possessive of that person,' Luther continues. 'You become fearful they may leave you for someone else; you become suspicious, then angry and maybe violent. Yes, love has its dark side.'

'Okaaay', Jennifer puffs out her cheek. 'Anyone want to add something?'

'It's depressing,' says a girl's voice.

'It's the way of Homo sapiens,' says Martin.

'Martin, you're a student of history,' says Jennifer. 'I suppose your dark definition of Love is based on ...'

'My studies? Yes, partly, though fortunately there is a ...

Bright side to love, yes, Luther?'

Luther nods and sits down, putting his legs up again, he's enjoying this.

'May I?' continues Martin, as this time he picks up the blue marker pen.

Jennifer puffs out her cheeks again. 'Be my guest.'

'Love is also Acceptance,' he says and starts to speak and write another list of twenty six words, parallel to the first list.

Acceptance
Brave
Caring
Dancing
Empathy
Forgiving
Gratitude
Honest
Inquisitive
Joyful
Kind
Laughter
Music
Nature
Observant
Patient
Quiet
Respectful
Singing
Tolerance
Unconditional
Valuing
Welcoming
eXciting

Youthful
Zestful.

Again he stands to one side and studies his list.
'...the Bright side of Love, and even more importantly ...'
This time Martin picks up the red marker pen and underneath the two lists he writes in big bold letters.

DARK OR BRIGHT?

WE HAVE THE POWER TO CHOOSE

Martin returns to his seat.
The students look at the board in silence.
Sam stands up with his I-pad, walks toward the white-board and takes a photo.
He nods at Martin who smiles and nods back.
Quietly, every student does the same.

So ... Love has a dark and a bright side, opposites again. Yin & Yang. Love & Hate.
In Chinese philosophy the Universe can only exist as a combination of Opposites. It is called The Way (Tao) and is immensely complex so can hardly expect our own existence to be any less complex.

With our awareness of being aware, and having responsibility for our actions, we are now co-creators of The Way, whether we like it or not ... by our actions, deeds, thoughts and words.

We are no longer innocent.

Everything we say, do and think has an impact on the Tao – on the stream of Universal Consciousness – whatever we wish to call it.

It's a tough call for we don't just flow with The Way, as

do other life-forms – all governed by the Yin Yang forces of nature which regulate supply and need, maintaining a level of harmony which just happens to have created a beautiful planet.

We evolved, by some extraordinary means and by our own consciousness, to become The Way.

We can interrupt that Flow, (all consciousness can), often disastrously – by our own decisions; its okay to make mistakes but not if wisdom and understanding fail to emerge.

Our growth in wisdom and understanding is painfully slow and not everyone can agree on what is wise or good.

There is a yard-stick, however, that if recognised and put into practice means we can stop making the same mistakes over and over.

It is the one step that can reverse the most damaging loop of our existence ... the Loop of Lethal Violence against our own species.

Without Lethal violence the chance for Homo sapiens to evolve into a race of Homo spiritus becomes a real possibility. A human that can share; that is no longer afraid of not having enough and who can rejoice in the rich variety of human cultures and races without suspicion, fear or bias.

This loop of Lethal Violence is driven by a Force that knocks the wind out of us.

That force is called Love yet it is not the same kind of love that a mother cat feels for its kittens.

Oh deary me, no!

As humans developed a consciousness of the Self and became questioning and deeply aware of being aware, our feelings of love intensified, becoming a driving force in the behaviour between man and woman, of man against man, creating extreme emotions of jealousy, rage and hatred – the dark side of love.

History, archaeology and anthropology prove that modern humans have always suffered lethal violence inflicted upon each other due to these conflicting forces.

The struggle to find and keep peace has for ever been our nightmare.

We long ago ceased to behave like the Zebra – unconcerned and un-emotional about life or death.

There is no 'thought' of tomorrow for the Zebra but Humans plan constantly for the future, anticipating danger, anticipating loss.

So, how on earth do we solve this dilemma? Well, that last sentence holds a key, a key that could unlock our loop of lethal violence. It's been mentioned lots of times:

Fear.

It's now time to take a deeper look at Fear and how it rules our lives and our emotions and more importantly how it destroys Love.

Fear is all about 'Loss & Pain'

Apparently, babies are born with just two fears – that of falling and that of loud noises.

All other fears we 'learn' by experiencing them or witnessing them. If you get bitten by a dog you may learn to be afraid of dogs; if as a toddler, in a pushchair, you see your mum reacting with fear to dogs you too will learn this fear.

Clearly, for a baby, evolution has programmed it to fear falling because this means losing contact with its parent (usually the mother).

Loud noises often mean danger - attack, animals, earthquakes etc and therefore danger.

As fragile Beings on a planet hurtling through space and time at the mercy of gravity and challenging weather, we have so much to lose – our homes, our possessions, our livelihoods, our health and safety, our loved ones, our youth, our beauty, our sanity and of course, ultimately our lives.

We die.

All Life dies.

And because we become Conscious of this fact, death for many people becomes a terrible fear. We understand how vulnerable we are and that death can happen anytime, that' loss' can happen at any time.

We grow to be more afraid.

(Other animals have no concept of death as a future event.)

As human-kind developed a deeper Consciousness we began to be conscious of being conscious of suffering fear, pain and loss, we started to bury our dead and develop complex rituals and myths surrounding birth and death.

We began to anticipate 'loss' so started to develop resources (farming for example) to prepare for winter, for starvation, even for death – look at the Pyramids.

But by then something terrible started to happen, as Fear took a stronger, almost permanent hold in our Consciousness we started to be afraid of not having enough resources.

So what happened? We land-grabbed, we stole people for slavery, women and the wealth and resources accumulated by our 'neighbours'.

With the fear of losing everything we became highly competitive. (see part 1)

The man who could gather wealth and resources, or defences against attack, became a Leader of Men and that status gave him a sense of power and prestige he would fight to retain at any cost, even killing his fellow countrymen, and so psychopathic behaviour crept into the bio-emotional mix.

History is peppered with unspeakable cruelty inflicted by Leaders fearful of losing their Power, wealth and privilege.

Individual citizens too can turn into thieves, murderers and psychopaths, creating the very death, misery and fear we all wish to avoid.

And so a constant state of warring developed between families, tribes, clans, neighbours and then nations.

Anxiety became strongly coded into our genetic make-up and no wonder! We even became capable of feeling anxious for no particular reason. Hyper-anxiety or chronic anxiety is a common condition for modern humans.

And so it was that revenge, anger and hatred took hold deep in our hearts and the dark side of love flourished into something we call Evil.

And the same is still true today.

We still live with the fear of running-out of resources – be it food, oil, money, land, jobs.

We still accumulate weapons for defence or attack.

We still nurture bitterness and the desire for revenge from events that happened in the past.

Leaders still dream of Conquest and Empires, of endless wealth and power, no matter the misery and human suffering it causes.

Fear of losing what we have gained, or of not having enough, creates the very deprivation we fear most.

We all know it is madness.

We all know it needs to change.

So, what about Pain? Loss and pain go hand in hand. That much is obvious.

Pain

All animals feel pain, particularly those with a brain and a central nervous system.

For humans pain is two-fold, we have not only a

biological response to pain but a very strong emotional response too, made greater by our awareness of being aware of pain.

So, of course, we fear pain, even to anticipate pain is a terrifying experience.

And then we have pure emotional pain – the kind that destroys us when we *lose* someone we love. Grief and sadness can be unbearable. Even to imagine losing someone and the grief this would cause can make a parent to over-protective or a lover over-possessive.

It's the same with depression. In fact, depression is so widespread and prevalent among human beings it makes you wonder, it certainly makes me wonder .... could it be that depression too has somehow become encoded into our genetic make-up?

And does that surprise us? Good grief – it's tough enough being alive if you're an animal governed by instinct, but for humanity, being alive means making choices and look where that has got us.

The point is this, every single one of us is currently the 'end product' of a long, long ancestry of violence and hardship. We all of us have ancestors who caused or suffered immeasurable suffering due to the loop-de-loops of human behaviour.

Except for a rare few, life is undoubtedly a struggle so no wonder just being alive can cause depression and yet it is hardly to our advantage to feel this way so how come evolution has allowed it to happen?

Let's take a closer look ...

Depression

We will never know when depression first emerged as part of the human condition though it's possible to imagine it did so around the time we made that transition from trees to savannah to cave dwelling.

Maybe it went something like this:

As our original habitat of deep equatorial forests shrunk as the Earth grew warmer and grasslands took over; our ancient hominid ancestors were forced to leave the safety of the trees and their abundant food supply for the danger and uncertainty of these wide open savannahs.

At this point we had not developed speech, or consciousness of being conscious and were still governed by our instincts for survival. But, oh boy! Did we have to start thinking on our feet – literally!

It would have taken generations and generations of adaptations and learning, trial and error and yes, decision making, and all the while our brains grew larger and our diets changed as we were forced to eat meat. We became hunters and foragers - dangerous activities, taking us away from the relative safety of our clan and requiring new skills.

And so we learnt to use sticks, stones and bones as tools, and our thumb responded to this evolutionary challenge by shifting its position so tool making became easier and more refined.

Communication too, first in the form of hand-signs, developed as hunting and foraging became essential for our survival. We have always lived in groups and this didn't change, in fact, it became essential for our survival, a male or a female on their own would not have survived long.

All these challenges, as well as being more exposed to weather systems and the prey of other animals and the scarcity of water and competition from other groups of struggling hominids, made life fearful and treacherous.

Females and children 'learned' to scream to tell their clan members they were in mortal danger. (We still have this fear-response today, even if it's just a little girl being chased by her brother.)

Young males had to acquire new skills for hunting and defence. The open sky (a mystery in itself, particularly the movement of stars, moon and sun) the clouds and weather ... foot-prints and animal tracks ... gave clues as to what to expect.

We started to become aware of having a future; of anticipating trouble; our bonds became stronger with those we felt safe with and we began to feel the loss and pain of death. When a hunting party of males left, for maybe hours, the females learnt to be afraid they would not all return and there was always the fear of a failed hunt, of starvation and the death of infants and children.

Post-traumatic stress would have been as relevant then as it is today but how does evolution cope with this?

Even before we developed a complex language of spoken sounds, humans needed to express their feelings, just as other animals and primates do – snarling, roaring, thumping the ground; or stroking, grooming, cuddling and playing.

So what is the 'language' for post-traumatic stress?

Body language, that's what.

The last thing any hunting or foraging party needed was a member who was 'not up to it'. Whatever the cause, if a female or male, huddled into a foetal position, rocked, keened, refused to eat, refused to leave the cave then the message was 'leave me alone. I can't function. I'd fail you, I'd put the rest of you in danger.'

This behaviour must somehow have been tolerated and may even have led to sympathy and ultimately to that side of humanity that is caring and compassionate, further strengthening depression as an evolutionary adaptation for survival of the clan.

So the thing is this ... it's perfectly normal to feel depressed. It's been bred into us but there's no denying it is a huge blight on millions of lives making any kind of life-fulfilling achievement massively more difficult.

The next question is ... what does Homo spiritus do about it?

First we need to look at what depression is; and what it isn't; based on my work as a holistic therapist, and then we may find a way to change it.

- Depression is not an emotion; it is a state of mind, a very painful, frightening state of mind.
- Depression is not the same as grief or sadness, or feeling blue or fed-up or miserable, these are all emotions that, if left un-checked, can develop into low-level depression.
- Depression can be mild, moderate, acute, chronic, life-damaging, including thoughts of suicide. It can come and go in any strength, unexpectedly and often unwarranted.
- There is often a desire for oblivion; a sense of extreme hopelessness; the feeling of tipping over and into a large black hole with no way out; the complete and utter absence of joy.
- A feeling of uselessness and that 'friends and family would be better off without me'.
- A feeling of unworthiness, of being unlikeable and then pretending to be jolly.
- An incomplete sense of identity.

Phew! That's a pretty bleak outline of a common human condition and if you tick any points in that list, trust me; you're not alone, not by a long shot. The most bewildering fact for most sufferers is that they can't explain why they have this terrible state of mind.

Picture the scene ...

It is 3.11 in the morning. The bedroom is dark; outside, the world is silent. In two minutes time Carol will start to wake up; she won't know why and 45 seconds later she will recognise she is awake and at 3.15 a.m. she will feel herself sliding into a deep dark ocean of utter despair.

This happens a lot, since she was about fourteen years old and Carol cannot understand it. She lies in bed, next to her husband who is sleeping soundly. Carol feels her muscles tighten, her breathing becoming short and shallow, she wills herself to return quickly back to sleep before the waves wash over her head and drag her down into the depths of gloom.

A dry sob catches in her throat. As always she wants to cry but the tears never flow.

'For god's sake!' she says to herself, knowing that sleep has vanished from her mind.

'For god's sake!' she repeats as her eyes open and she stares into the darkness of the room and listens to the soft breathing of her husband. She wants to scream, but never does.

She wants to know why she has this sense of doom, of hopelessness and the dread of a new day.

The worst of it is Carol suffers alone for she has never been to the doctors, never told her husband or her daughter, never told a living soul. How can she? When she couldn't begin to explain why she feels like she never wants to wake-up.

'I have a wonderful husband, a married daughter who has just given us a beautiful baby grandson. I like being a dental receptionist, have a great boss, no serious money or

health problems so why on earth should I feel so depressed?'

The other thing is Carol knows that when she gets up in the morning she will put on a cheerful face, go about her usual routine and by the time she has had breakfast and gone to work the depression will have lifted.

Occasionally, a wave of hopelessness will wash over her but by keeping herself busy her mind doesn't dwell on it.

Still, her deepest darkest fear is that her desire for oblivion may one day prove too strong.

Today, that desire is overwhelming so she gets out of bed ... her coping strategy is to go into the kitchen and make a cup of hot milk and add a dash of brandy, she doesn't like the taste much but sometimes it helps her get back to sleep.

On the fridge is one of Suzi's cards, held in place by a fridge magnet. Carol studies it and sees there is a website. Five minutes later she's logged-on and is looking at a single page website; it gives the days and times of classes and a photo of a young woman called Suzi and a middle-aged man called Martin; there is a list of workshops and a list of problems that may be helped, insomnia and depression were included. The page stressed that these workshops were not therapy groups and were for self-healing and well-being only.

There was a contact e mail...

Carol sent a message, requesting if there were any spaces for a Tuesday evening.

So, recognising depression is part of your personality, as it is for millions of other humans and then deciding to get help are two important steps for dealing with the problem.

It is worth noting that hormones and other chemicals play a huge part in this evolutionary development – the two go hand-in hand; as do various life-cycles such as puberty, child-birth, old age, even the cycles of the moon and the seasons. Seasonal Adjustment Depression - SAD is well recognised by medical establishments.

Homo sapiens have gained a huge amount of knowledge and help for clinical depression; the following are some examples:

- Share and talk about your depression.
- See your doctor – blood tests may show a deficiency in Vitamin D, for example.
- Listen to your doctor's advice and ask about any alternative methods before you start a long course of anti-depressants. Cognitive Behaviour Therapy and other forms of counselling can be very helpful.
- Find a reputable, recommended Holistic Therapist who can work alongside your doctor and explore other avenues such as acupuncture or aromatherapy.
- Exercise gives a sense of purpose and achievement releasing hormones that make us feel better.
- Caring for something – a pet, has the same affect.
- Take note of any cycles in depression, for example with P.M.T or the phase of the moon – the full moon is usually the worst time; then at least we have an understanding and know the phase will pass.
- See a nutritionist for advice on diet or do some

research. We are what we eat.
- Apathy and lethargy are all part and parcel of depression. Seek small goals.

From a Homo spiritus view-point, Clinical Depression – the kind we have no control over, as with Carol, is the last stage in the journey for Homo sapiens.

When someone makes that ultimate choice ... that ultimate and last act of free-will – to consciously and deliberately ends their own life by their own hand then the Dark side of Love has won the day and nobody is to blame.

Maybe there just comes a time when we can take no more, as individuals and a species.

The Dark side of Love is the ink that has written our history books, a history full of suffering.

There is even the thought that if enough people feel the desire for oblivion then this is what we will achieve; the belief that we are doomed and not deserving of our planet home gives some people the idea that we don't deserve to survive.

Hmm.

A scenario of mass-suicide?

But should we be deciding for our children and their children? Don't they deserve at least the chance to put things right and don't we have a duty to start that process?

Homo spiritus could be that process and ... it's not too late.

What we have to do is start reversing the love-hate loop.

Turning Hate to love

Tragically, the human fear response to pain has for eons been a weapon in the hands of Power - using torture; threats against our loved ones, horrendous methods of execution and extreme deprivation to subjugate and control populations.

The truth is Evolution has programmed Homo-sapiens to guard, at all costs, against the threat of loss and pain. That cost has been our peace of mind, our sense of joy.

We are so wired to become angry and to feel hatred and to seek revenge it is the only way we know how to 'put right a wrong'.

An eye for eye

Vendettas

Honour killings

Land-grabbing

War

It is time to change; if we don't then our extinction by our own hand is ever more likely.

Yet, if we are hard-wired to act so aggressively, with fear and anger and hatred; to seek revenge and conquests; to desire to have the 'most', how can we change?

By using our will-power, that's how, and by becoming conscious of when our emotions pull us towards the Dark side of Love and then deliberately reversing that Loop.

But is Martin right? Can we honestly make a choice between the dark and the bright? If someone has broken our heart can we honestly chose to feel acceptance rather than anger? And if that someone finds a new lover can we honestly choose not to feel jealous but joy for their new happiness?

The truth is we simply don't know how to Love Well.

We need to ask ourselves why love so easily turns to hate.

What are we afraid of?

Well, my darlings, when it comes to the human heart we are afraid of so many things, all to do with loss and emotional pain – Abandonment, Isolation, Rejection …

Our need for Love is extraordinary, almost out of proportion.

Some would say it is the basis for our existence, that from the moment we are born we search for, seek and need all that Love can give us – security, wholeness, peace, joy, a sense of belonging, of being loved, of being admired and praised.

The fear and pain of losing all that can change the very core of our being. It's as though being loved gives us our identity.

The truth is love and hate are two sides of the same coin that flips over quicker than you can say dynamite …

Picture the scene...

Kate is in the canteen at Uni, having lunch with her friends. Her phone rings – she glances at the name – it's Phil, she smiles and waves her phone at the others before dancing outside to take the call in private.

Kelly watches her, 'bet that's her boyfriend, she's sooo in-love with him.'

'Yeah, he's coming up this weekend,' said Sam.

They watch her and then everything changes.

Kate starts to pace, furiously, back and forth; she drags her fingers through her hair, pulling at the roots. Her face is crumpling, she starts to yell, to cry and then she stares at the phone. It has gone dark, it has gone silent.

In her melt-down Kate hurls the phone into some shrubs, covers her face with her hands then breaks into a run, away from the canteen.

'Oh my god!' Kelly says and jumping to her feet runs after her friend.

The others stare, in shock.

Martin is sitting, reading, under a tree nearby, he has heard and seen Kate's grief; seen her throw the phone and run off down the path. Getting up he goes to the bushes and searches for the phone.

Sam stands up. 'I know that guy, his name's Martin, look he's found her phone.'

The force of love we feel in our hearts is like a fire that can blaze out of control and become destructive or it flickers like a weak flame, afraid of its shadow or simply dies and smoulders in its own embers.

This is not The Way.

The Cosmos (the Totality of Everything) strives always for harmony. Cycles will continue to loop; the pendulum will continue to swing until the extremes of too much Yin and too much Yang are in balance.

Beauty and peace is harmony; harmony is beauty and peace. The Way of Nature reflects this and how we love should reflect this also.

Up to the age of seven children accept and give love without any conscious awareness, if they are surrounded by love they flow with it, like a bubbling brook.

Between seven and fourteen we start to become more aware of our own identities and aware of being aware; we start to become reflective and love starts creating a whole new landscape. We have started to experience loss and pain and friendships become complex.

At puberty love flows with such force we are often out of our depth, we have learnt to manipulate and always there is the fear of rejection.

So, Love is like a river whose source is a trickle in some secret place where it begins its journey to the sea and on its way it becomes a brook and then meets up with other brooks and streams until a mighty river is formed, transforming and nourishing the landscape.

There is little more beautiful and peaceful on this planet than a river valley; who has not stood in awe at the scenery and the sounds of nature in this place of perfect harmony?

However, when the river is in flood (too Yang) it destroys the landscape and when the river bed is dry and empty (too Yin) it cannot support the life it previously nourished and becomes a wasteland until the rains come again.

Homo sapiens are also caught up in a swing of extremes –

Domination/subjugation
Fear/anger
Loss/gain
Too little/ too much

Until we know how to Love Well, it cannot change.

Until we consciously decide to change, change cannot happen and we will continue to suffer loss, pain and depression.

And yet there is hope for although we are victims of a violent history, we have an understanding of what harmony is – it is beauty which is perfection and we try to capture it in Art, music and the written word.

It is always about proportions – about form – about content and context; ask any musician, writer or artist or designer. Not until the balance is perfect will they be satisfied with their work. Every single note of music counts.

The problem is, we just can't quite achieve this harmony in our relationships with one another, with other nations and races. It is our tragedy.

In the Realms of Consciousness harmony is experienced as Love in its purest form without the dark side or the side of sentimentality and attachment.

Deep within our own Consciousness we know this and yet our ancestral bio-emotional loops keep us chained to the dark side of love, and to break free we need to Reverse the Loop of love-hate.

And by god, it's not easy.

This will be the hardest thing we will ever do, breathing correctly is a doddle compared to this, but breathing correctly was an essential first step.

To choose the Bright side of Love we need to take a leap like we've never taken before. It is a leap across the Abyss and that takes enormous courage not least because we are blind-fold.

This is something very few have dared to try and being blind-fold we have no idea what we meet on the other side, or even if we can reach that far.

While courage is our strength it is faith that carries us.

Faith is a word of mystery. It is both belief and conviction and yet it is more than that or the very word itself would not exist.

Faith means to trust and then to surrender to that trust.

Faith

Okay, now we're getting close to the nitty-gritty of what Homo spiritus is all about.

Firstly, we are co-creators of our own reality.

We are not just part of The Tao, we *are* The Tao, we made that leap millennia ago. We have access to The Field of All Possibilities, recognised even in Science as a plausible dimension.

The point is this, when a life-form develops consciousness and a free will then those possibilities becomes endless.

You may like to read that last bit again.

We have no idea how powerful we truly are, not the kind of power to impress or dominate or control, a different kind of power all together, one that can achieve what we all long for - Peace.

It is a power harnessed by the Bright side of Love, a power that can change our genetic code, to de-program our knee-jerk reactions of fear, greed, anger and hate for only then can we truly experience peace as a human condition.

A Peace that will give us huge opportunity to truly evolve into an extraordinary Being, not the kind obsessed with naked ambition for personal gain but for something we can't yet even begin to imagine.

Secondly, if fear is the lock that keeps us chained to the loops of destruction, Faith is the key that will unlock them.

Have faith that those twenty six words of the Bright side of Love can work a miracle ... trust in the power of their meaning and their intention.

And then let go of that Intention, let go, surrender and

let that power flow. Words like acceptance, caring, forgiving, laughter, patience, tolerance ... only words and yet they hold such power!

So, where do we start with this faith ... this trust that such words can bring about change?

Well, we start with you and me. It's an old saying but it's perfectly true ... if you can't love yourself, then how can you love anyone else?

To be honest this is a lot easier than it sounds. Most people, people like Carol, can't get their heads round the idea of loving themselves. It sounds a bit narcissistic, as though you think you're a bit precious.

Actually, it's exactly what Carol needed to do. Until she could learn to accept herself exactly as she was, depression and all; until she could really care that she existed and forgive herself and her ancestors and all humanity for creating the D.N.A that was shared with millions; until she could learn to be patient and tolerant towards herself, then change was never going to happen.

To achieve all this, Carol was asked to look in the mirror and really gaze into her own eyes, deeper and deeper and say 'I love you'. Like others in the workshop she could only hold her own gaze for a fleeting second or two. It just didn't feel ... right!

Eventually, when she could hold her own gaze quite comfortably she was asked to smile at her reflection with total non-judgement. To release her shoulders, breathe from her tummy and smile into her own eyes.

Just smile, with faith that the magic will work.

And that is when the healing starts, with a smile and by not making any judgements. Self-loathing is often unrecognised. Love is the only cure and it comes from those 26 magical words.

By following the six steps of Homo spiritus, Carol was

able to cure her insomnia and her bouts of depression. Nine nights out of ten. On that odd night she'd accept the black dog trotting behind her, turn, pat him on the head and said 'come along, then, just a little way'. Then smile herself back to sleep.

So much for one person

Still puzzled as to how society can change?

Picture the scene...

Someone is crying behind a closed door. Sam and Martin walk softly along the hall towards the door. Sam knocks gently. Kelly opens the door a crack and peeps out.

'Hi,' she whispers.

'Hi,' whispers Sam. 'Remember Martin?'

Kelly nods.

' Well, he's found Kate's phone ... he wonders if he can help ...?'

'I hope someone can.' Kelly says and opens the door. 'Kate, Sam is here with Martin, that guy I told you about.'

'Is it okay if I come in?' says Martin.

Kate sits up, she's curious. 'Okay.'

She wipes her face with the back of her hands; her eyes are red and pinched.

Martin walks into the room; pulls out the chair from the desk and sits down in front of Kate. 'I have your phone, I saw you throw it into the bushes.'

Martin puts the phone on the bed, next to her. Kate stares at it, her breathing quickens. Sam and Kelly stand in the doorway, watching.

'I thought you might want it back,' says Martin. 'You'll probably want to talk to Phil at some point. I couldn't help

hearing what you said.'

Kate gulps and sniffs. 'I never want to speak to him again, I hate him, he ...'

She starts crying again.

Martin turned to the other two. 'Any chance of a cuppa?'

'Sure,' says Sam and he races off, quite glad to escape, to be honest.

Kelly sits next to Kate and puts an arm across her shoulder. 'Did Phil say why he wasn't coming up this weekend?'

'He said his car's got an oil leak but that's just an excuse ... it's his bloody rugby, it always is. Anyway he could've got a train, like I do but he's too bloody tight, unless it's rugby and drinking with his mates, he ...'

Kate breaks down again and covers her face with her hands.

'You haven't broken up, have you?' says Kelly. 'You two have been together for ages! Kate, don't let this one ...'

'I told him how upset and angry I felt, Kelly and he said he couldn't take it and how difficult long distance relationships are and ... so I said did he want to break up with me and he said he needed to think about it, the bastard! Just wait till I go on face book and tell everyone how harsh and selfish he really is!'

She grabs the phone and through her tears starts stabbing at the buttons.

'Before you do that,' says Martin. 'Will you do something for me?'

Kate looks up at the older man with his kind face and gentle manner. 'What?' She sniffs.

'... In return for me grubbing about in the bushes, to find your phone?' He adds.

Kate gives a half smile. 'Okay.'

Sam comes back with a tray of steaming mugs, the room

is very quiet; he stands and watches as Martin gently takes the phone from Kate's grasp.

'Will you just rest your hands, like this?''

Gently Martin takes hold of Kate's wrists and lays them on her lap, palms face-up. His touch is warm and very kind, Kate starts crying again, silent tears.

'Uncurl your fingers, Kate and let your hands and wrists relax completely.'

Holding her wrists Martin waits for this to happen. 'Now close your eyes, unclench your teeth and try a tiny smile ... I know it's hard, you don't feel like smiling but ... to please me.'

It was a brave smile, a difficult smile.

Sam and Kelly watch in silence.

'Now, Kate, will you hold Phil's face in your mind's eye, a face you remember with love and happiness.'

Kate sniffs and gives small violent shakes of her head.

'Just keep trying, smile through the pain.'

'I can't!'

'Just try.'

Kate's breathing shudders and she stops crying. Her wrists slacken; her breathing shudders again and her head droops a little.

Martin lets go of her wrists.

'Kate,' he continues; his voice soft and almost hypnotic. 'You are going to work a miracle ...'

Kate lifts her head and stares at him, the others stare at him. Sam puts the tray down, spell-bound.

Martin says. 'Trust me, Kate, I know how this works. Nothing, ever, happens in isolation. Okay?'

She gives a tiny nod, closes her eyes and relaxes once again.

'Continue to see Phil in your mind's eye and with your beautiful inner smile tell him 'I love you, I'm sorry, forgive me.'

Kate's eyes snap open, she's shocked and angry. 'No way! He's the one who should be sorry; he's the one who …'

Martin puts his finger to his lips. 'Shh. Nothing happens in isolation, Kate.'

She stares at him, not understanding.

'What do you mean by … a miracle?' *Sam asks.*

Martin looks at them all. 'Well, the power of unconditional love can work miracles, If Kate's love for a fellow human being, in this case, Phil, is not conditional upon him staying with her, even of him loving her, then anything can happen.'

'Such as …?' *Sam asks.*

'Will he come back to me?' *Kate says.*

'Who knows?' *Martin lifts his arms and grins.* 'The possibilities are endless. Maybe tomorrow, Kate, you will talk to a boy you've not looked at twice; maybe Phil will concentrate on his Rugby and make a successful career of it? Maybe, one day in the future you two find each other again? But in the long term, it will be for the greater good; that is the miracle of the unconditional love.'

Kate puffs out her cheeks and looks doubtful.

'Let's try again. Close your eyes, breathe, relax, think of Phil, smile and say to him "I love you, I'm sorry, forgive me." Say it more than once, until it becomes easy.'

Kate sighs and closing her eyes, she follows Martin's instructions.

Sam and Kelly watch in fascination as Kate's face becomes a landscape of tiny muscle movements, betraying a host of emotions. Finally her face is calm; she takes a very deep breath, opens her eyes and stares as if in a trance.

'How do you feel, Kate?' *Sam asks.*

She blinks at him. Her voice is quiet. 'Just sad.'

'And the anger and hatred?' *Martin asks.*

Kate gives a tight smile. 'Gone.'

'Wow.' Breathes Kelly.

'When the anger and hatred return,' says Martin. 'Repeat what you've just achieved and all will be well. The sadness is natural and will go in time. Okay?'

Kate looks at him and nods her head. 'Thank you,'' She whispers.

He smiles at her. 'Here's your phone and if you can't say anything nice, which would be mega powerful, say nothing; which is almost as powerful.'

He chuckles and leaning forward kisses Kate on her forehead. 'Remember ... I'm sorry, I love you, forgive me. Keep saying it till you believe it, in your heart and mind.'

Kate nods.

Sam looks at Martin. 'Can I ask something?'

'Sure.'

'You're pretty cool, for an older guy, I mean ...' Sam blushes to the roots of his dark ginger hair. 'I mean, what makes you so calm and ... stuff.'

Martin smiles. 'Lots of things, for me it all started with T'ai Chi. Heard of it?'

Sam nods. 'Yeah, there's classes advertised on the notice board.'

'That's me; you're welcome to come along, all of you.'

Standing up Martin picks up a mug of tea and walks to the door. He turns and smiles at them.

'We can't change the pass, but we can heal it.'

The thing is this, if we all adopted this reaction to those who have hurt us, in the near or even distant past, then the miracle spreads outwards. It can envelope the planet.

Wow!

What more is there to say? It's tempting to leave it here, with Martin walking out of the door and Kate staring at her phone, deciding what to text Phil, if anything; with Sam and Kelly sipping their tea, watching her, waiting.

It's tempting to leave everything here; to see if the bigger picture emerges on its own; to see if these six steps can help lead us to a new horizon, but the jelly is set so now we can turn it out and study its shape, form and colour.

Do we recognise it?

It has been shown to us before, several times but for the most part we've ignored it due to ignorance because we've been blinded by our own bio-emotional loops of loss and pain.

We can, if we chose, remove the blind-fold but first we have to take that leap of faith that if we dare and trust to love our enemies (never forgetting we are also their enemy) that this love and trust is reciprocated then ... well, in time there will be no more enemies.

This may seem like a fairy tale, but wait ... we have yet to taste the jelly because we don't know its flavour.

How will we do this?

Very carefully, jelly is wobbly stuff ...

This is when we take a closer look at our DNA and the havoc it can cause, DNA that has been bred into us due to the Dark side of Love being repeated over and over again throughout our long and violent past.

Because if we don't understand how 'wickedness' evolved then how can we ever heal more than just a girl's broken heart?

Ready?

Our D.N.A

There is a branch of science called Evolutionary Biology which studies how animal behaviour, including our own, is programmed by the challenges in our environment.

Generally, evolution favours adaptations that ensure the survival of the species. As we have seen, depression is one such adaptation but it is not the only one, not by a long shot.

Human behaviour is so very complex and it is tragic and mystifying to observe the many Personality Disorders that have been programmed into our genetic code. Many of which do us harm and make our lives so much more difficult and even dangerous.

The question is how did this happen? If these traits are so catastrophic for our survival how did they become coded into our DNA?

It doesn't make sense until we remember that WE HAVE CREATED OUR OWN ENVIRONMENT.

Therefore the answer lies in Society because it is the Environment of Human Society with our wars and greed and fear; distrust and revenge that have created so much anger, paranoia and cruelty.

Unbelievable suffering and cruelty has been inflicted in all manners of ways and circumstances, generation after generation, is it any wonder psychotic behaviour became encoded into our very cells?

Society means you and me, and all of your ancestors and all of mine. When humans live and interact in close proximity nothing happens in isolation. In fact nothing in the Universe happens in isolation. Homo spiritus knows this.

The thing is this - no one is blameless and those of us who consider ourselves sane and normal may do so by sheer luck and good fortune.

It is easy to judge a person by their behaviour and point the finger and say 'he is evil', 'she is a monster', or 'he's weird', 'she's neurotic'.

What we forget is that Society has created these kinds of people and it is likely that we all have a tendency to behave in some form of uncontrolled manner, it all depends on triggers and where on the spectrum we happen to be.

People with OCD, (Obsessive Compulsive Disorder) while not harmful to others, can have a difficult life and how difficult depends on where they are on the spectrum. Just a little way may result in 'counting'; at the extreme end they are likely to live in a lonely sterile 'prison'.

The point is this, no one chooses, as a baby, to grow up and have 'weird' behaviour or do cruel things – it is already in their genetic make-up and if the circumstances lend themselves then those genes will 'activate'.

Someone with the genes for alcohol dependency (a stress-release mechanism) will never become alcoholic if they never have an intoxicating drink in their life.

The Dark side of Love flourishes because our genetic code has adapted to our violent natures, over and over again. It is the chicken and the egg scenario, which came first? The violent behaviour or the violent genes?

It doesn't matter, they happened side by side and until we can change our behaviour those genes will not be bred out of us.

Or – until these genes have been bred out of us we cannot change our behaviour.

This is the Loop that requires nothing short of a miracle.

How can someone with, for example, narcissistic sociopathic genes, possibly change their natures?

They can't.

Okay, time to take a spoonful of jelly ...

The most instinctive thing we do in any relationship ... with ourselves, our family, neighbours, friends, strangers, foreigners, enemies ... is to make judgements.

We do it constantly, we walk down the street, we sit in coffee shops, we watch the news, we look in the mirror and make instant judgements based on what our eyes tell us and what we believe is good or bad, right or wrong, ugly or beautiful.

We point the finger and make sweeping statements.

Okay, this is normal human behaviour. We need to make judgements, to be discerning so we can avoid the wrong company or a situation that may put us in danger.

We need to be able to make choices and no-one in their right mind would suggest anyone took into their homes or their hearts someone with sociopathic or psychopathic tendencies.

So, the jelly you are about to swallow is not the kind of jelly you are used to; the Love you are about to discover is not the kind of love you are used to giving and receiving.

In order for Homo sapiens to heal their violent past two things must happen.

First.

Switch off the Ego

Nothing is personal, not ever.

If a man beats his wife it is not her as an individual ... it would be any woman who was married to him. If a mother neglects her child it would be any child under the same circumstances.

Phil did not set out to break Kate's heart, in that sense it wasn't personal either, even if he had done so for some sadistic kind of pleasure, it would still not have been personal – he would have been responding to his genetic make-up that prompted him to behave so cruelly and Kate wouldn't be the first or the last girl he treated so badly.

If a country invades another country that too is not personal – it is all the past leading up to that opportunity and that time and place and that Leader who takes that decision. Not one citizen could say – 'this is about me'

By seeing the bigger picture we can switch off the Ego and our Self-centred opinion and judgment of someone, we can then take a truly OBJECTIVE viewpoint.

By achieving an objective 'judgement' we can then see that it is not the person but the deed that is evil.

The person is the catalyst in an ever-ending scenario of cause and effect stretching back throughout our history, way back to our earliest ancestors who began to be aware of being aware and taking actions on their own account, by their own free will, no longer acting for the survival of the clan but for their own desires.

So when it comes to blame, blaming parents or grandparents for our own particular nightmares who does that generation blame for theirs?

Only when we cease to point the finger and blame, can we hope to change our catalytic behaviour and to do this we need to change our mind-set from one of blaming to one of acceptance.

Accept that things are as they are because of our natures and set about changing our natures to change the way things are.

You might like to read that sentence again.

The way to reverse that loop is with the power of Bright Love. Look at that list again. Really look at it. See how many of those attributes you incorporate into your everyday life and relationships.

Martin's advice of 'I love you, I'm sorry, forgive me' is the key that can unlock us from our chains of cruelty, revenge and hatred.

Here's a true story...

There was once a very brave lady (I think she was Dutch) who worked as a courier for the Resistance against the Nazis in World War 2. On one assignment she got into a serious spot of bother and ended up hauled before a Commandant who demanded to see her papers. Her story was weak, her bicycle had broken down, she looked dishevelled and her papers were dubious.

As he scrutinised her and her papers she forced herself to relax and in her heart she said, 'God bless the Christ in him.'

She didn't know where the words came from but the effect was immediate. The Commandant sighed, folded the papers, gave them back to her and told her to go.

For this particular lady, Christ was the embodiment of Love. For other people it could be Buddha or Shiva or Krishna ...

Love is the key but it is not the kind of love that is about 'me'; it's about everyone else.

It's not a question of who has done wrong ... all our ancestors have done wrong, societies create the Dark side of Love and we are all of us 'society'. No one is outside it.

Society creates the madmen; the deviants and the perverts. These people did not choose to inherit those genes.

Of course, those who are a danger to the public need to be removed from society, that goes without saying. Prisons and various forms of punishment are as much a part of human society as anything else but they fail catastrophically to change the Dark side of Love.

If anything our judicial systems perpetrate fear and hate, bitterness and revenge. Forgiveness and understanding is unheard of, except on very rare occasions.

But then if someone murders your child how is it possible to understand let alone forgive such an act?

This is the crunch of the matter; this is when you decide you're brave enough to taste the jelly.

Taking that Leap

This is going to be short and sweet, or not.

Be prepared for a knee-jerk reaction, this is when you taste the jelly and you may not like it.

The very next time someone bugs you or you feel resentment, dislike, loathing or even hatred for somebody, be it someone you know, maybe someone you once loved or a complete stranger or someone in the news, sport, entertainment, politics ...

Maybe it's the face of a terrorist who is being hunted or has been caught for some terrible atrocity ...

Maybe it's the face of someone in the past who has caused unspeakable suffering – the Hitlers, the Stalins, the Pol Pots ...

Become aware of your emotions of the Dark side of Love and how your chest seems to fill with such loathing ... take a deep breath, relax, keep breathing and with an objective mind 'smile' at the image of that person and say in your heart, 'I love you, I'm sorry, forgive me'.

So, it doesn't taste nice, we feel like spitting it out; it seems like poison, a betrayal to all those who suffered and still suffer but REMEMBER this does not have to be personal; when healing the violent past of humanity we are 'speaking' for all of eternity, for all our ancestors and future generations.

The Field of All Possibilities is not bound to space or time.

Consciousness – individual, universal or Cosmic, is not bound to space or Time

Love can heal the past, the present and the future which in the material world is only a breath away.

When you first start practicing this technique of Reversing the Loop of love/ hatred it will feel like a lie. Your heart and mind are in conflict, you will struggle to make even the tiniest of smiles, to breathe calmly and to say those words with any kind of conviction.

There will be days that you forget and get caught up with your darker emotions; there will be many times you try to reverse the loop and fail; I know this.

The point is to keep trying.

The point is to guard your thoughts, words and deeds. When you realise your gossiping a little too unkindly about someone, stop and undo the damage.

When you realise you've made a judgement about a homeless person lying in a doorway, or a drunkard, or a morbidly fat person, stop that thought and undo the damage.

When you start to manipulate someone or engage in tit-for-tat on face book ... stop and reconsider your motives.

If your motives come from that list of Dark Love then know it is time to switch to the Bright.

It will be the hardest thing you ever do; I know this.

But put it this way – none of us are perfect, we have all hurt someone who has been left with bitter and angry memories, so wouldn't it be extraordinary and humbling to think that that person, recognising that nothing happens in isolation asks *us* in their heart for forgiveness and sends unconditional love our way.

This is at the heart of cultures that believe in Karma and

reincarnation. With the realisation that no-one is blameless, a light then switches and we then understand we all need to ask for forgiveness and that this is the key that will heal the past for individuals and humanity as whole.

All we need is the Intention and the will to try - the will to transform our natures.

It will take time, practice, repetition, humbleness and determination, but we can do it; we can all do it.

Over time, and maybe quicker than we imagine, the knock-on effect multiplies until that Critical Mass is reached.

A Critical Mass is a term used in science and social dynamics. It is the smallest amount of something needed to start and then maintain a chain reaction, a reaction that then becomes self-sustaining and creates further growth.

In the fields of consciousness this Universal Law applies to all nature.

Here's a true story...

There was once a little bird, a blue-tit that lived among humans and flitted about minding his own business, keeping out of the way and doing what blue-tits do.

One of these things was to be attracted to anything that glittered or shone in the sunlight. Being of a curious nature he would examine such things.

One day, as a young bird he saw, in the early morning sunlight something shining on the front doorstep of a house.

It was the metallic-foil top of a milk bottle. Being particularly inquisitive and brave he flew down to investigate, perched on the edge of the bottle and tapped at the foil with his beak.

Imagine his reaction when his beak pierced the foil and

encountered the rich cream floating on top of the milk.

Now this was a taste he discovered he liked! So, guess what? Being intelligent, as is all life, he quickly learnt to keep a beady-eye out for the milkman doing his round and before long he was enjoying a daily sip of cream.
Blue-tits are flock birds and it didn't take long for all his brothers and sisters to see what he was up to and to start copying. Very soon all the house-owners in the road, who had their milk delivered on their doorsteps every morning, discovered holes in the milk tops and wondered what the milkman was up-to.

Okay, this doesn't seem particularly amazing but what happened next was astonishing. Over a period of time this blue-tit behaviour became common in the local area, as you can imagine, but then this behaviour 'exploded' and blue-tits were observed country wide pecking holes in milk bottle tops.

Far quicker than simply spreading from one locality to the other.

This is an example of Critical Mass, a sufficient number of blue-tits learned this behaviour for it to suddenly become a chain reaction that then became self-sustaining and created further growth.

It as though this behaviour became 'imprinted' in blue-tit consciousness and so spread out across time and space.

The same is true of human consciousness and is how our mind-sets change.

When sufficient people start reversing their ancient bio-emotional loops and start using the Bright side of Love instead of the Dark then a miracle can happen.

Maybe not in my life time, or yours but the sooner we

make a start the better. It takes just you and me to get the ball rolling.

In our own quiet way.

The way of Homo spiritus.

One last thing, before we go our separate way, there is one word in Martin's list of Bright Love that we need to take a good look at ...

Gratitude

If the Dark side of Love is the lock that imprisons us and faith is the key that turns it then it is Gratitude that pushes the door wide open, letting all the Brightness of unconditional Love flow into our lives.

We all know it is polite to say thank you. It is when we stop saying thank you, through idleness or carelessness that the source of all we need dries up and stops flowing into our lives.

Here's another true story...

Derek was a pretty ancient kind of guy, well into his eighties, never married but had at least dozen nephews and nieces, great nieces and great nephews.

He'd been a coal miner then a postman; worked hard all his life, saved hard, had a couple of pensions, tended his allotment and grew vegetables for himself and his neighbours.

One winter's day he came to see me and he was decidedly grumpy. Was it the cold? The weather? His aches and pains?

No.

He was unhappy because out of all his relatives to whom he sent cheques for Christmas, only three had written to get in touch to thank him.

'If the others can't even be bothered to tell me the money had arrived safely I don't feel like sending them anymore, ever again,' he said.

And who could blame him.

Taking things for granted is never a good idea, although we all do it.

What is a good idea is to pause now and again, maybe last thing at night before settling down to sleep, or when stuck in a traffic jam or sitting on a train, staring out of the window and think of all the things we should feel grateful for.

It's not just Great Uncles but the Cosmos likes to hear from us, too.

The Realms of Consciousness are very powerful Realms and should not be underestimated.

Haven't we all sent out pleas for help in times of crisis but do we always remember to send a note of thanks?

The truth is, the nature of the Cosmos is one of give and take, Yin and Yang, remember? It's never a question of punishment but one of balance and flow.

Everything is about cause and effect - that is the engine that drives the Cosmos. Nothing happens in isolation.

So, when something bad happens we curse our bad luck; we're quick to blame others and pour scorn, derision, anger and maybe hatred on those we think are responsible ... pouring all that negative emotional energy out into the ether, building up the force of Dark Love.

Adversity is always a reminder that something is out of sync. And oh boy! Haven't we been out of sync forever and a day? Isn't that what our violent past is all about?

Gratitude is an astonishingly powerful medicine that can heal not only the past but also the present and the future.

Picture the scene...

Dark clouds are gathering, a summer storm is brewing; the wind picks up and the city prepares itself for a lashing.

Curling through the centre of the city is a canal, left over from the age of steam, and recently restored for pleasure trips and barge holidays.

Martin wakes up to the sound of rain beating on the roof a few feet above his head. He is renting a small barge for his three years at Uni; it suits him perfectly.

The rain does not.

He pulls aside the curtain and frowns at the clouds and the curtain of rain. The tow-path will be a track of mud. No good for cycling on. He'll have to walk to Uni and it's a good hour's trek.

He'd better get moving.

Going to the stove to put on the kettle, he sees an envelope lying on the mat by the door. Picking it up, Martin opens it and finds a letter informing him that the mooring fees are going to increase at the end of the month.

His frown deepens.

Sitting down with his mug of tea and a bowl of muesli and banana, Martin recognises his dark mood. He puts down the mug, rests his hands on his lap, closes his eyes and regulates his breathing; there is something he needs to do.

Martin needs to fix this mood before he walks out the door or the day could get worse.

He thinks about the things that are bugging him then starts a prayer of gratitude.

'I am grateful for the rain, grateful I have clean running

water on tap; that I don't need to dig a well or sanitise my water. I am grateful for the rivers and canals, for living in such a beautiful green land. Rain is a gift and I am grateful for all the gifts this planet home provides.

I am grateful for this canal; and the people who keep it clean and safe and the opportunity to live on this barge and call it my home.

I am grateful for my chance to study and all my tutors and new friends; I am grateful for the struggles that come my way for they will make me stronger and more considerate.

I am grateful for all that I have and for every experience that comes my way.'

Opening his eyes, Mikes takes a deeper breath, sits still for a moment, recognises his mood has lifted, smiles and picks up his mug of tea.

The day did get better; on the way to Uni he met someone he hadn't seen for a long time and they had time for a quick coffee and a catch-up.

Walking home later that day he passed the moorings office, popped his head and said he'd received the notice and was then told it was sent in error because the length of his barge was below the stipulated size.

Sometimes it is unbelievably hard to change a negative experience into something positive, let alone give thanks for it but when we do, be prepared for the unexpected.

This is something Homo spiritus try to do.

It can change the world.

Gratitude and the Bright side of Love go hand in hand. In America they have Thanksgiving Day; perhaps every day should be a Thanksgiving Day – for our food, our friends and maybe even our enemies who will then cease to be our enemies.

This is the miracle that is waiting to happen.
So, let's finish this journey with a party!

At the restaurant where Luther works there is a loft which can be hired for private parties, a long wooden table on a wooden floor boards, rustic beams, candles and balloons.

Kate is here, sitting quietly. It is the weekend that Phil was due but she hasn't heard from him and she didn't send a text or put anything on face book; she accepts it's over.

Kelly and Ellie, Sam and Mike sit nearby, chatting. Luther is on duty and is waiting-on, a big grin on his face.

It is Suzi's birthday and she thinks she is meeting a few friends for tequilas and nachos. She is standing across the road, waiting for a break in the traffic as a double-decker bus goes by.

A teenager is sitting on the top deck; he sees the purple hair, catches his breath and leaps to his feet as Suzi darts across the road and into the doorway of the restaurant.

The teenager leaps off the bus, the driver shouts at him. The boy ignores him and runs up the street.

Upstairs in the Loft, Suzi is amazed to see so many friends as well as students from her classes. Carol is there as is Ellie and her mum, who has also joined. They all clap and cheer.

Mike gets up and walks towards Suzi and kisses her on the cheek.

'Happy birthday Suzi, glad you could make it.'
'And I suppose this is your doing?'
'With a bit of help from the others.'
They sit down at the table.
'How do you two know each other?' *Ellie asks.*
Suzi leans forward on her elbows. 'You know that angel I mentioned?'

Ellie nods.

'Well, this is my angel; he rescued me from myself a long time ago.'

'You took my help,' says Martin. 'Besides what goes around comes around, I was a fallen angel once, too, remember?'

Luther appears and puts drinks on the table.

'This is my poison, now,' says Martin. 'A large tonic with ice and lemon.'

Ellie stares at him and nods her head then turns to Suzi.

'I've got some news. I've applied to do a degree in Dietetics. I'm going to be a dietician!'

'Brilliant!'

One of the waitresses is running up the stairs. 'Luther, we've got a problem, can you come?'

The others fall silent and watch as Luther strides away. Five minutes later he's back with a young man in tow.

'Suzi, this is Glen, he says you know him … he says you've met.'

Suzi stands up, she stares at the young man intently, there's something familiar about him.

'Yeah, it is you,' says the young man. 'You do remember me don't you? I was that bloke you …err … in the Mall a few months back. You … err …gave me some dosh.'

Suzi's face brightened. 'Of course I do. How are you?'

I'm okay, it's just, well, I've been looking out for you cos I've got my shit together, err .. sorry …'

Everyone laughs.

'That's okay Glen. It's my birthday, sit down and join us and tell me what you've been up to.'

Luther pulls up another chair and everyone shifts along.

Kate's phone beeps a text. She glances at it and sucks in air.

'What?' says Kelly.

Kate reads it then turns to her friend, confused. 'It's Phil, wanting to know where I am.'

'Cheeky bugger! Unless ... tell him, you've nothing to lose.'

Fifteen minutes later a taxi pulls out in front of the restaurant and once again Luther is called downstairs. He returns, grinning.

'Any room for one more gate-crasher?'

Suzi looks up, 'guess so. Who is it?'

Luther beckons and a tall, broad young man walks into the Loft.

Kate squeals.

Kelly whoops.

Phil blushes.

Sam gets up and pulls Phil towards his own chair, next to Kate who is now speechless.

'Sit down, buddy! I'll get you a pint.'

Sitting down, Phil turns to Kate and cups her face in his hands. 'I've been an idiot. I love you. I'm sorry, can you forgive me?'

The past cannot be changed but it can be healed.

***

Footnote...

If you'd like to read more about Universal Consciousness, The Totality of Everything, the Field of all Possibilities and other things mentioned in this book ... and things that haven't, such as our higher-conscious mind ... you may like to read Three Worlds – a book to be published in the near future.

If you've enjoyed the short stories you may like to read a trilogy called The Phoenix and The Dragon – a fictional account of two gifted teenager Homo spiritus, girl-boy twins - who discover they have the potential to change the course of humanity and have to battle against not only those who wish to destroy them but also their own doubts and fears. To be published a little later in the future.

Lastly, if you'd like practical experiences of this book then why not attend one of our residential Homo spiritus workshops in Scarborough, North Yorkshire, England. Please e mail info@fountainscourt.com and we will be happy to send you details.

Recommended Reading

The truth is; the author is not a scholar; her memory for facts and figures is poor. However she is a lapsed member of Mensa and for her, knowledge and learning happens when a 'light' is switched on, illuminating something that she can understand as a 'picture', something that her imagination can get to grips with and yet appears to be logical.

Her quest for knowledge started at a very early age. Here is a list of some of the books and authors that influenced her thinking and helped, from those childhood days, to create her vision of the world, the Universe and the Cosmos. Most are not too scholastic and can be enjoyed by anyone, of any age with a curious mind.

- *I-Spy books*
- *Books by Romany of the BBC*
- *Any books by Lyall Watson, Linda Goodman, Deepak Chokra, Stephen Hawking, Allan Butler, Jostein Gaardner.*
- *The writings of Dr Edward Bach.*
- *The Four Agreements by Don Miguel Ruiz*
- *The New Testament.*
- *The Tao Te Ching*

Recommended Therapies

As a therapist for many years the author has first-hand knowledge of alternative therapies and modalities, some ancient, some new.

Any of these, alone or as a combination can help anyone fledging Homo spiritus achieve their goals:

- Bach Remedies (Consultation+ remedies))
- Yoga
- T'ai Chi & Chi Gung
- Emotional Freedom Technique (EFT)
- NLP (Neurolinguistic programming)
- Life Coaching
- Meditation
- Clinical Hypnotherapy.

Above all else, have fun and enjoy your journey.

Printed in Great Britain
by Amazon